MIRL

OSSICLE MORPHOLOGY OF SOME
RECENT ASTEROIDS AND
DESCRIPTION OF SOME
WEST AMERICAN FOSSIL ASTEROIDS

BY

DANIEL BRYAN BLAKE

UNIVERSITY OF CALIFORNIA PUBLICATIONS IN GEOLOGICAL SCIENCES

Volume 104

UNIVERSITY OF CALIFORNIA PRESS

OSSICLE MORPHOLOGY OF SOME RECENT ASTEROIDS AND DESCRIPTION OF SOME WEST AMERICAN FOSSIL ASTEROIDS

OSSICLE MORPHOLOGY OF SOME RECENT ASTEROIDS AND DESCRIPTION OF SOME WEST AMERICAN FOSSIL ASTEROIDS

BY

DANIEL BRYAN BLAKE

UNIVERSITY OF CALIFORNIA PRESS
BERKELEY · LOS ANGELES · LONDON
1973

WILLIAM MADISON RANDALL LIBRARY UNC AT WILMINGTON

University of California Publications in Geological Sciences

Advisory Editors: W. A. Clemens, G. H. Curtis, A. E. J. Engel, R. M. Kleinpell, J. H. Langenheim, M. A. Murphy, R. L. Shreve, C. A. Wahrhaftig

Volume 104

Approved for publication June 16, 1972
Issued August 15, 1973

University of California Press
Berkeley and Los Angeles
California

❖

University of California Press, Ltd.
London, England

ISBN: 0-520-09472-7

Library of Congress Catalog Card No.: 72-87674

© 1973 by the Regents of the University of California

Printed in the United States of America

QE783
.A7
.B55

CONTENTS

270288

OSSICLE MORPHOLOGY OF SOME RECENT ASTEROIDS AND DESCRIPTION OF SOME WEST AMERICAN FOSSIL ASTEROIDS

BY

DANIEL BRYAN BLAKE

(Contribution from the University of California Museum of Paleontology)

ABSTRACT

ANALYSIS OF thirty species representing eleven genera and four families of Recent asteroids suggest that the morphology of discrete skeletal ossicles is sufficiently consistent intraspecifically and variable interspecifically to be useful in taxonomic studies. Ontogenetic variation appears to be less than interspecific variation. The ossicles of seventeen of eighteen Recent species of the genus *Luidia* closely reflect their systematic arrangement as interpreted by Döderlein (1920) and others but indicate that *L. foliolata* Grube should be assigned to the Alternata Group instead of the Clathrata Group. The morphological differences between *Platasterias* and some species of *Luidia* are less than the differences among species assigned to *Luidia*. *Platasterias* is therefore assigned to the genus *Luidia*, family Luidiiae, of the subclass Asteroidea.

Comparative morphological studies of adambulacral ossicles of representatives of eight families suggest that these ossicles have been conservative in evolution. Structures considered homologous can be found throughout these families. Evolution also appears to have been conservative in the ambulacral ossicles of the four families considered. Marginal ossicle morphology is more diverse, but similar morphologies occur in members of the Astropectinidae and Luidiidae, suggesting homologies.

Luidia etchegoinensis, n. sp.; *L. sanjoaquinensis*, n. sp.; *Mistia spinosa*, n. gen., n. sp.; *Nehalemia delicata*, n. gen., n. sp.; and *Sucia suavis*, n. gen., n. sp., are described from west American Cretaceous and Cenozoic rocks. Although fossil data are limited, there is no evidence of basic change in the nature of the west American starfish fauna during the Cenozoic.

INTRODUCTION

ALTHOUGH MANY papers have been written describing more or less complete fossil asteroids, individual ossicles generally have not been emphasized unless entire specimens were unavailable. Very useful papers on individual fossil ossicles have been published (Müller, 1953; Hess, 1955), whereas Viguier (1878) studied skeletal morphology of Recent asteroids. Studies on ossicle morphology are more common in other echinoderm groups, for example, the work of Murkami (1963) on ophiuroids. It is still true, as Hess (1955) and Howe (1942) pointed out, that morphology of individual ossicles is usually neglected. The purposes of the present paper are to evaluate the taxonomic usefulness of individual elements and to interpret possible evolutionary relationships in light of data derived from ossicle morphology. An attempt has been made to devise a consistent format for ossicle description.

The asteroid skeleton consists of many unfused (except for some in the mouth region) ossicles, covered by an epidermis and more or less deeply buried in a dermal layer. The ambulacral areas are restricted to the oral surface. They are in the form of a groove, the arch of the groove being formed by the paired ambulacral ossicles. The adambulacral ossicles support the ambulacral ossicles and form the margins of the groove. Specialized ossicles are present in the oral region. In many asteroids, including most of those considered here, relatively large ossicles, termed marginals, form the margins of the arm. They are usually found in two series, the superomarginals above and the inferomarginals, on which they rest. The remainder of the arms and disc are covered by many, usually closely articulated ossicles of variable morphology. A great many large and small spines, tubercles, and spinelets are attached to most principal skeletal elements.

The rarity of fossil asteroids is probably largely due to their unfused skeletons; upon death, the tissue usually decomposes rapidly, and the ossicles are disassociated. Further, the rather open meshwork of the typical asteroid ossicle may be more vulnerable to physical and chemical weathering than are the massive skeletons of some other echinoderms.

In order to study morphology of ossicles, it is necessary to completely remove the soft tissue. This is most rapidly done in sodium hypochlorite (common household bleach). In the specimens prepared, the bleach did no apparent damage to the ossicles in the relatively brief period of immersion necessary for the destruction of the soft tissues. Illustrated ossicles were dipped in red ink, mounted on plastic stubs, coated with ammonium chloride, and photographed.

A severe limiting factor in the study is the mutilation of specimens needed to obtain the ossicles for study. Specimens of many species are very rare; it was necessary to damage those specimens as little as possible. It was also often necessary to avoid excessive damage to specimens of species that are relatively abundant; hence, in every case ossicles of only a limited number of specimens were studied. Ossicles taken from medial and proximal portions of arms of mature, but generally not unusually large specimens were emphasized. Largely because of the damage problem, ossicles of the oral area have been studied in only a few species. The writer therefore considers all descriptions to be subject to revision in the light of broader sampling of larger suites of specimens.

Acknowledgments

I am deeply indebted to Dr. J. Wyatt Durham, under whose direction this study was undertaken, for advice and encouragement, and to him and Miss Maureen E. Downey for discussions and critical reading of the manuscript. I am very grateful to Dr. David L. Pawson and Miss Maureen E. Downey (U. S. National Museum) and Dr. Charles R. Stasek (formerly of the California Academy of Sciences) for their generous help in allowing my use of facilities and specimens under their care; and to Dr. Donald P. Abbott (Hopkins Marine Station), Dr. John S. Garth (Allan Hancock Foundation), and Mr. Jack G. Vedder (U. S. Geological Survey) for permission to study specimens belonging to their institutions.

I have also benefited from discussions with Drs. William B. N. Berry, Cadet H. Hand, and Edwin C. Allison.

Support for illustration was provided by the National Science Foundation

Graduate Subvention Funds at the University of California; a Faculty Summer Fellowship from the University of Illinois permitted study at the U. S. National Museum.

GLOSSARY

Technical terms and phrases used in this study are defined below. The list is not intended as a complete glossary of asteroid terminology. Detailed descriptions of ossicles have necessitated the proposal of new terms or descriptive phrases, but the usage of other workers is followed as much as possible. Important sources include Spencer and Wright (1966); Fisher (especially 1911); Hess (1955); Rasmussen (1950); Schuchert (1915); Spencer (1914–1940); Verrill (1914); and especially Müller (1953). Capitalized abbreviations, such as "N" are after Müller. These abbreviations are followed by Müller's German-language descriptive phrase. In order to refer to a structure type that is present on both proximal and distal surfaces of an ossicle, the prefixes "p" and "d" have been applied to the Müller terms; for example, pN refers to the proximal N.

An ossicle or taxonomic name enclosed in parentheses following the term signifies that the usage of the term is restricted to that ossicle and/or taxon.

The letters and numbers following a definition refer to the abbreviation used for the structure and a text figure on which the structure is illustrated. For example, *aboral ridge* is abbreviated *abrg* and is illustrated in text figure 1G. Orientation information at the ends of definitions refers to the associated drawings.

Problems of consistency of usage arise in attempting to apply the same terms to ossicles and to the complete animal, and to taxa with acute interbrachial angles and to those with broad interbrachial angles.

The long axis of an ossicle commonly does not parallel the "length" of the arm on which it is found. When referring to an individual ossicle or ossicle structure, *length, width,* and *height,* as well as their qualifying adjectives (*long, short, wide, narrow, high, low*) are used only in the same sense that they are used on the entire arm, not with respect to the dimensions of the ossicle or individual structure referred to. Thus, the "length" of an ossicle is that dimension which is parallel to the "length" of the arm on which it is found, even if the "width" is the greater dimension. Structures on ossicles also are described with respect to arm orientation; the "length" of a structure on an ossicle parallels the length of the arm and the width is perpendicular to the length and lies in the horizontal plane. For example, a slender, cylindrical structure whose long axis is vertical and which is found on the adradial face of an ossicle would be called *high* and *short,* not *long* and *narrow.* Terms other than *length, width,* and *height* may be used without regard to orientation; a structure on any face or in any position could be called *elongate* or *slender and cylindrical.*

In a taxon with acute interbrachial angles, such as *Astropecten,* the Amb, Adamb, InfM, and SupM series are approximately parallel and radiate from the oral area. In taxa with broad interbrachial angles, the Ambb and Adambb still radiate from the oral area, but the marginals do not parallel the Ambb and Adambb series throughout. For example, in *Ceramaster,* the "length" of the ossicle is parallel to the "length" of the arm distally, but as the marginal series diverges from the arm axis in the proximal direction, so does the "length" of

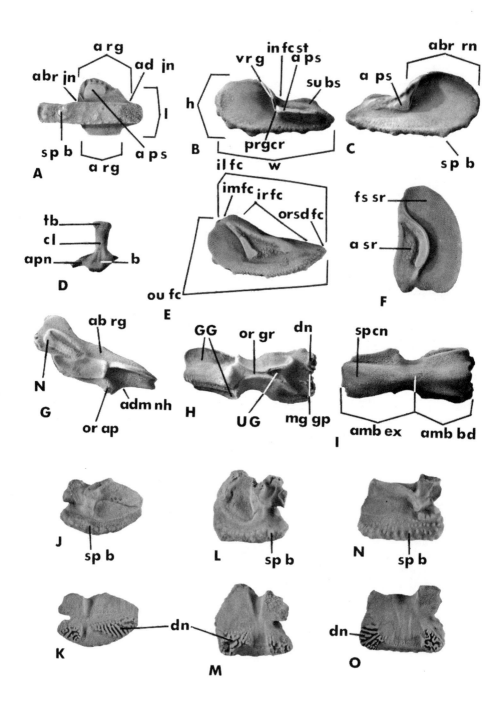

the ossicle diverge from parallel to the "length" of the arm. In order to be consistent in usage, all terminology is applied to all taxa as it is applied to the acuteangled taxa. In *Ceramaster* and in taxa of similar shape, ossicle length is that dimension along the line of successive marginals between arm tips, and not a dimension measured parallel to the arm axis and the Amb-Adamb series.

aboral—That surface opposite the surface bearing the mouth, or that portion of a structure toward the aboral surface.

aboral ridge (Amb)—The keel-like aboral surface extending across the width of the ossicle (abrg, 1G).

abradial—The direction away from the arm axis.
abradial apophyse (Amb)—In certain starfish, the abradial Adamb articulation structure (pl. 18, fig. 29).

abradial constriction (InfM)—A constriction in the outline of the oral surface near the abradial margin of the ossicle (pl. 6, fig. 16) (oral view, adradial right).

abradial junction (InfM, Luidiidae)—The abradial contact between the distal articulation ridge and the main body of the ossicle (abr jn, 1A) (oral view, adradial right).

abradial region (InfM)—The relatively high part of the ossicle away from the arm axis (abr rn, 1C) (adradial view, adradial left).

Fig. 1. Illustration of morphological features of InfMM, SupMM, Ambb, Pax, and mouth ossicles.

A–C, *Luidia foliolata* Grube, InfMM, ×6; A, oral view, hypotype 10647h; B, proximal view, hypotype 10647f; C, distal view, hypotype no. 10647g.

D, *Luidia foliolata* Grube, Pax, ×6, hypotype no. 10647a.

E, *Astropecten armatus* Gray, InfM, lateral view, ×4, hypotype no. 10669j.

F, *Astropecten armatus* Gray, SupM, lateral view, ×4, hypotype no. 10669r.

G–I, *Luidia penangensis* de Loriol, Ambb, ×9; G, lateral view hypotype no. 10646i; H, oral view, I, aboral view, both hypotype no. 10646g.

J, K, *Luidia (Platasterias) latiradiata* (Gray), mouth ossicle, ×6; J, laterally directed face, hypotype no. 10634k; K, face directed toward opposite member of mouth ossicle pair, hypotype 106341.

L, M, *Luidia phargma* Clark, mouth ossicle, ×6; L, laterally directed face, hypotype no. 10645i; M, face directed toward opposite member of mouth ossicle pair, hypotype no. 10645h.

N, O, *Astropecten armatus* Gray, mouth ossicles, ×4; N, laterally directed face, hypotype no. 10669s; O, face directed toward opposite member of mouth ossicle pair, hypotype no. 10669t.

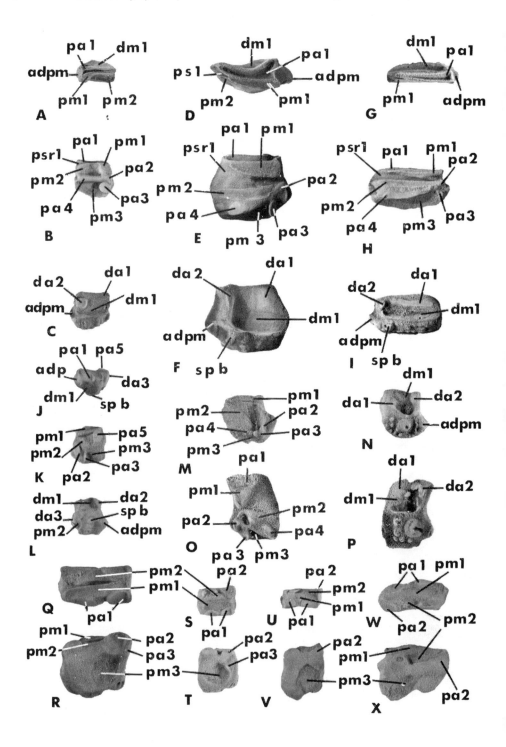

adamb notch (Amb)—The oral abradial surface that articulates with the Adamb (adm nh, 1G) (lateral view, adradial right).

adambulacral—Ossicle of the series that borders the ambulacral furrow and supports the ambulacral ossicle (*abbrev.* Adamb, pl. Adambb). Orientation of these ossicles is somewhat variable. In the Luidiidae, Astropectinidae, and Benthopectinidae they are imbricate, are inclined distally, and are idealized in the descriptions as two-dimensional objects possessing proximal and distal surfaces only. Goniasteridae Adambb are foreshortened and have distinct aboral and oral surfaces.

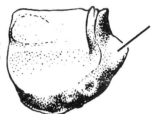

ad pm (Adamb)—The more-or-less angular *adradial prominence* that extends into the ambulacral furrow; it serves as a guide for the tube foot (ad pm, 2F) (oblique distal view).

adradial—The direction toward the arm axis.
adradial apophyse (Amb)—In certain starfish, the adradial Adamb articulation structure (pl. 18, fig. 29).

adradial junction (InfM, Luidiidae)—The adradial contact between the distal articulation ridge and the main body of the ossicle. The ridge may be notched at the contact (ad jn, 1A) (oral view, adradial right).

Fig. 2. Morphological features of Adambb and comparative morphology of Adambb.

A–C, *Astropecten armatus* Gray, ×4; A, oblique aboral view, ambulacral furrow left, hypotype no. 10669b; B, oblique proximal view, ambulacral furrow right, hypotype no. 10669d; C, oblique distal view, ambulacral furrow left, hypotype no. 10669c.

D–F, *Luidia foliolata* Grube, hypotype no. 10647b, ×6; D, oblique aboral view, ambulacral furrow right; E, oblique proximal view, ambulacral furrow right; F, oblique distal view, ambulacral furrow left.

G–I, *Luidia (Platasterias) latiradiata* (gray), ×9; G, oblique aboral view, ambulacral furrow right, hypotype no. 10634d; H, oblique proximal view, ambulacral furrow right, hypotype no. 10634m; I, oblique distal view, ambulacral furrow left, hypotype no. 10634d.

J–L, *Archaster typicus* Müller and Troschel, ×9; J, oblique aboral view, ambulacral furrow left, hypotype no. 10681c; K, oblique proximal view, ambulacral furrow left, hypotype no. 10681b; L, oblique distal view, ambulacral furrow right, hypotype no. 10681a.

M, N, *Cheiraster gazellae* Studer, ×6 M, oblique proximal view, ambulacral furrow right, hypotype no. 10673n; N, oblique distal view, ambulacral furrow right, hypotype no. 10673o.

O, P, *Luidia elegans* Perrier, hypotype no. 10649b, ×9; O, oblique proximal view, ambulacral furrow left; P, oblique distal view, ambulacral furrow right.

Q, R, *Nidorellia armata* (Gray), ×6; Q, aboral view, ambulacral furrow right, proximal up, hypotype no. 10682b; R, proximal view, ambulacral furrow right, hypotype no. 10682a.

S, T, *Mediaster aequalis* Stimpson, ×6; S, aboral view, ambulacral furrow right, proximal up, hypotype no. 10680a; T, proximal view, ambulacral furrow right, hypotype no. 10680c.

U, V, *Dermasterias imbricata* (Grube), ×6; U, aboral view, ambulacral furrow right, proximal up, hypotype no. 10683b; V, proximal view, ambulacral furrow right, hypotype no. 10683a.

W, X, *Patiria miniata* (Brandt), ×6; W, aboral view, ambulacral furrow left, proximal down, hypotype no. 10684b; X, proximal view, ambulacral furrow right, hypotype no. 10684a.

amb body (Amb)—Adradial part of the ossicle; the amb body and amb extension together from the complete ossicle (amb bd, 1I) (Ambulacral körper) (aboral view, adradial right).

amb extension (Amb)—Adradial generally elongate part of the ossicle, not sharply differentiated from the amb body. The amb body and amb extension together form the complete ossicle (amb ext, 1I) (Ambulacralfortsatz) (aboral view, adradial right).

ambulacral—Ossicle of the double series that forms the arch of the ambulacral furrow and articulates with the Adamb (abbrev.: Amb, pl. Ambb).

articulation process—A subcircular contact and articulation structure. It is usually raised and of denser skeletal meshwork than the surrounding ossicle material and contacts a similar structure on an adjacent ossicle (a ps, 1A).

articulation projection (Pax)—Elongate process that extends laterally from the base and serves to link adjacent paxillae (a pn, 1D).

articulation ridge—An elongate enlarged contact and articulation structure, not all of which may be in close contact with the neighboring ossicle (a rg, 1A) (lateral view, adradial right).

articulation surface—That area of the face including and enclosed by the articulation ridges (a sr, 1F).

base (Pax)—Enlarged oral terminus of ossicle (b, 1D).

column (Pax)—Elongate slender or swollen stalk between base and tabula of ossicle (c, 1D).

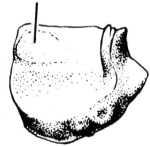

da1 (Adamb, *distal face articulation surface 1*—A frequently inconspicuous surface near the aboral margin of the distal face; this surface bears against the pa4 of the next distal Adamb (2F) (oblique distal view, adradial right).

da2 (Adamb, *d*istal face *a*rticulation surface *2*)—A usually inconspicuous process located medially along the adradial margin on the distal face; this process bears against the pa3 of the next distal Adamb (2F) (oblique distal view, adradial right).

da3 (Adamb, *Archaster*, *d*istal articulation surface *3*)—A prominent abradial process that contacts the adjacent InfM (2L).

dentition (Amb)—Adradial cuneate or tabular structures that articulate with corresponding structures on the other member of the amb pair (dn, 1H) (lateral view, adradial left).

dGG (Amb)—Distal GG.
distal—Any direction (or area of an ossicle) away from the oral area that lies in the plane of the oral surface or in a plane parallel to the plane of the oral surface.
dN (Amb)—Distal N.

dm1 (Admb, *d*istal face *m*uscle *1*)—Depression for muscle that links successive adambb (2F) (M of Müller, see pm3) (oblique distal view, adradial right).

dm2 (Adamb, *Archaster*, *d*istal face *m*uscle depression *2*)—Depression for muscle that links adamb to InfM (2L).
face—Any planar or curved surface of an ossicle.

fasciolar surface (marginal)—The area on the side face other than the articulation area, largely occupied by the fasciolar spinelet bases (f s, 1F) (lateral view, adradial right).

first paxilla (*Luidia*)—Paxilla of first series aboral to InfMM in position of SupMM of other genera (*abbrev.* Fst Pax). The relationship between Fst Pax and SupMM is considered in the discussion section "Comparisons Between Members of the Luidiidae and Astropectinidae."

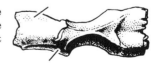

GG (Amb)—Winglike structures near the abradial end of the aboral margins; for attachment of muscles that connect the Amb to the Adamb (GG, 1H) (Flügel zur Verbingung mit den Adambulacralia) (oral view, adradial right).

granule—A more-or-less equidimensional, secondary ossicle found on the outer face of a primary ossicle or imbedded in the surface tissue.

height—The perpendicular distance between the idealized plane of the oral surface and the most distant point of the aboral surface; or any comparable interval used in reference to an ossicle or ossicle structure (h, 1B). ["Comparable" rather than "parallel" because maximum dimensions and symmetry planes and axes of ossicles are often not oriented parallel to the axes of the organism. In these cases, for descriptive purposes, the ossicle orientation is idealized. For example, in *Astropecten armatus,* the SupM are illustrated (pl. 14, figs. 54, 55) with their long axes vertical; in life, the long axis may be nearly-parallel to the height in proximal ossicles but form an acute angle in medial ossicles of the same arm.]

inferomarginal—Enlarged ossicle of the series that forms the oral lateral margins of certain starfish, oral to the superomarginals (abbrev. InfM; pl. InfMM).

inner face—Any interiorly directed ossicle surface that faces the body cavity rather than another ossicle (in fc, 1E).

inner face step (InfM, *Luidia*)—The break in curvature on the inner face (in fc st, 1B) (lateral view, adradial left).

internal face—Any interiorly directed ossicle surface; may face either another ossicle or the body cavity (il fc, 2E).

intermarginal face (marginal)—That face which contacts a marginal of the opposite series (im fc, 1E).

length—The distance between the center of the mouth and the arm tip (the distance R) measured parallel to the idealized plane of the oral surface, or any comparable interval used in reference to an ossicle or ossicle structure (see note under *height*) (1, 1A).

medial gap (Amb)—An interval, or break, frequently present between plates of the marginal dentition on an ossicle (md gp, 1H).

mouth ossicle—Enlarged ossicle located at the proximal end of the adambulacral column, paired with another mouth ossicle from the adjacent ambulacral column; the pair is directed toward the mouth.

N (Amb)—Surfaces on the proximal and distal sides of the amb body for articulation with neighboring Ambb; consists of an aboral muscle depression and an enlarged articulation process (N, 2G) (narbenförmige Grube zur Insertion kurzer dorsaler Längsmuskeln und Verbindung mit dem über stehenden Teil des adoral folgenden Ambulacrale) (lateral view, adradial left).

Oral—That surface with the mouth and ambulacral furrows, or that portion of a structure toward the oral surface.

oral apophyse (Amb)—In certain starfish, the Adamb articulation structure adradial to the Adamb notch (or ap, 1G) (lateral view, adradial right).

oral flange (Adamb)—An extension of the pm3 oral to the main body of the ossicle (pl. 1, fig. 6).

oral groove (Amb)—Groove on the oral surface extending abradially from the Unt G (or gr, 1H) (oral view, adradial right).

oral intermediate—Any of the primary ossicles between the adambulacrals and the inferomarginals.
oral side face (InfM)—Internal face that contacts the ossicles of the oral surface (or sd fc, 1E).
ossicle—Any individual skeletal element.
outer face—Any face of an ossicle that forms part of the body surface of the starfish; outer
  faces of primary ossicles are commonly at least partially covered by secondary ossicles (ou
  fc, 1E).

pa1 (Adamb, *p*roximal face *a*rticulation surface *1*)—A groove
  against which the Adamb notch of the ambulacral ossicle
  bears (2D, 2E) (oblique aboral view, adradial right).

pa2 (Adamb, *p*roximal face *a*rticulation surface *2*)—A process
  that contacts the distal side of the oral apophyse of the Amb
  whose dGG is connected to the pm2 (2E) (oblique proximal
  view, adradial left).

pa3 (Adamb, *p*roximal face *a*rticulation surface *3*)—A process
  that contacts the da2 of the next proximal Adamb (2E)
  (oblique proximal view, adradial left).

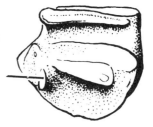

pa4 (Adamb, *p*roximal face *a*rticulation surface *4*)—A process
  that contacts the da1 of the next proximal Adamb (2E)
  (oblique proximal view, adradial left).

pa5 (Adamb, *Archaster*, proximal face *a*rticulation surface *5*)—A process that articulates with a process on the dGG of the Amb (2J, **K**).

paxilla—Columnar or hourglass-shaped aboral ossicle whose summit is variously covered with small spinelets or granules (pl., paxillae; *abbrev. sing. and pl.*, Pax).

pGG (Amb)—Proximal GG.

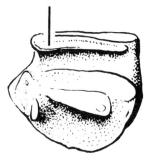

pm1 (Adamb, proximal face *m*uscle depression *1*)—A depression for the muscle that extends to the pGG of the Amb, whose amb notch articulates with the pa1 of the Adamb (2E) (oblique proximal view, adradial left).

pm2 (Adamb, proximal face *m*uscle depression *2*)—A depression for the muscle that extends to the dGG of the Amb proximal to the Amb linked to the pm1 and pa1 of that Adamb (2E) (oblique proximal view, adradial left).

pm3 (Adamb, proximal face *m*uscle depression *3*)—A depression for the muscle that links successive Adambb (2E) (*M* of Müller, Flache zum Ansatz starker Muskeln, über welche der Verbindung mit dem nächsten Adambulacrale erfolgt) (oblique proximal view, adradial left).

pN (Amb)—Proximal N.

primary ossicle—Any of the larger ossicles that form a major part of the test and typically are linked to more than one other larger ossicle.

proximal—Toward the oral area, in the plane of the oral surface, or in a plane parallel to the plane of the oral surface.

proximal articulation ridge corner (InfM *Luidia*)—The oral abradial angle of the proximal articulation ridge, visible in side face view (prg cr, 1B) (lateral view, adradial left).

psr1 (Adamb, proximal face *s*urface *1*)—A smooth surface between the pm1 and pm2 on the proximal face (2E) (oblique proximal view, adradial left).

secondary ossicle—Any of the smaller ossicles, granules, spines, or spinelets, many of which typically occur on a single larger ossicle or imbedded in the tissue, rather than in close contact with another ossicle.

side face—The lateral (proximally and distally) directed faces of an ossicle.

spine—An articulated slender or blunt ossicle attached to a larger ossicle.

spine base—Generally bosslike structure with which a spine articulates (sp b, 1A).

superambulacral—A small more-or-less cylindrical ossicle present in some starfish; extends from the Amb to the lateral ossicles.

superambulacral boss (marginal)—Prominence on the inner face that contacts the superambulacral ossicle (su bs, 1B).

superambulacral contact (Amb)—Area near the abradial margin of the aboral ridge against which the superambulacral ossicle bears (sp cn, 1I) (aboral view, adradial right).

superomarginal—Enlarged ossicle of the series that forms the aboral lateral margins of certain starfish; aboral to the inferomarginals (*abbrev.*, SupM, pl. SupMM).

tabula (Pax)—The swollen aboral terminus of the ossicle (tb, 1D).

Unt G (Amb)—Depression on the oral surface toward the adradial margin for muscle that connects the ossicle to the corresponding structure of the other member of the Amb pair (UG, 1H) (Grube zur Insertion des Unteren Quermuskels) (oral view, adradial right).

Vertical ridge (InfM, Luidiidae)—The vertically oriented component of the proximal articulation ridge (v rg, 1B) (lateral view, adradial left).

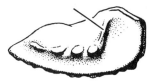

width—Any interval normal to the length in the idealized plane of the oral surface (see note under *height*) (w, 1B).

ABBREVIATIONS

CAS—California Academy of Sciences, San Francisco.

HMS—Hopkins Marine Station of Stanford University, Pacific Grove, California.

UCMP—University of California Museum of Paleontology, Berkeley.

USNM—United States National Museum, Washington, D.C.

SOME EARLIER STUDIES

Papers in the following list have been selected to illustrate some of the early, important, and varied studies on asteroid skeletons.

1732—Réaumur, R. A., was one of the first to make observations on asteroid hard parts; he observed that the lower surface of the skeleton was formed of many irregular trellisworklike pieces.

1842—Müller, J., and Troschel, F. H., considered the form, dimensions, and variation of ossicles and the position of papillae and pedicellariae.

1848—Forbes, E., recognized, described, and illustrated ossicles of fossil starfish. The number and form of ossicles were compared between genera.

1877—Agassiz, A., illustrated specimens with tissue partially removed and discussed ossicle relationships.

1878—Viguier, C., presented a classification based on the development of the ossicles around the peristome.

1891–1908—Sladen, W. P., and Spencer, W. K., described, illustrated, and compared ossicles of European Cretaceous forms in detail and suggested phylogenetic relationships.

1892—Perrier, E., related ossicle types in various genera and applied names to groups of ossicles.

1914–1940—Spencer, W. K., studied ossicles and evolutionary relationships.

1915—Schuchert, C., included ontogeny, regeneration, and evolutionary studies of ossicle systems.

1943—Nielsen, K. B., used marginal morphologies in generic descriptions and considered comparative taxonomic value of various ossicle types.

1950—Rasmussen, H. W., considered ossicle type and morphology in his study of European Cretaceous material.

1951—Spencer, W. K., classified large groups on the morphology of the ambulacral columns and discussed and illustrated ossicle morphology.

1953—Müller, A. H., ossicle types were considered individually, many descriptive morphological terms were applied, and different ossicle types were described in detail.

1955—Hess, H., applied morphological terms, discussed ossicle systems and general morphology, and included careful ossicle descriptions and detailed drawings.

1957—Lehmann, W. M., presented X-ray photographs of Paleozoic species.

1963—Fell, H. B., discussed asteroid phylogeny based on comparative morphology and growth patterns.

1965—Rasmussen, H. W., found starfish ossicles useful in stratigraphic correlations.

1965—Garcia, E. R., and H. H. Camacho illustrated some fossil ossicles and applied terminology in drawings of ossicles of a modern *Astropecten*.

1967—Heddle, D., described and illustrated starfish movement and relationships between hard parts and soft parts in species of *Luidia* and *Astropecten*.

1968—Weber, J. N., discussed carbon and oxygen isotope occurrences in the skeleton.

1969—Kesling, R. V., considered ossicle arrangement in starfish and revised the orders.

DISCUSSION

Introductory Remarks.—Asteroid ossicles reflect evolution in diverse ways. Among the various ossicle types, the Ambb and Adambb appear to have changed most slowly during evolution, because similar ossicle structures and shapes occur in taxonomically diverse genera. On the Adambb, analogous, seemingly homologous articulation structures can be found on all taxa considered here in spite of considerable differences in expression of these structures. There are morphological discontinuities, some larger than others, but these do not necessarily occur at major taxonomic divisions. Thus, Adambb of *Luidia,* of the Order Platyasterida, are basically similar to those of certain members of the Paxillosida Family Astropectinidae, but those of *Astropecten* itself are relatively distinct. Adambb of *Patiria miniata* (Brandt) and *Dermasterias imbricata* (Grube) of the Order Spinulosida are basically similar to those of *Mediaster aequalis* Stimpson and *Nidorellia armata* (Gray) of the Order Valvatida. If the classification does reflect broad evolutionary relationships between the luidiidae and the Oreasteridae, the Adambb became more massive and rectangular, the side faces relatively similar. These ossicles are strongly imbricate in the Luidiidae, but less so in presumably later families.

The Ambb, although not falling into more or less distinct groups in the manner of the Adambb, do appear to show some phylogenetic sequence of change in general shape. The adradial part of the Amb, termed the *Ambulacral body,* or *amb body,* is rectangular in some members of the Luidiidae and *Astropecten,* but massive and triangular or very long and essentially rectangular in other luidias and certain astropectinids and benthopectinids. The ossicle in general and the amb body in particular become shorter and somewhat imbricate and therefore asymmetrical. This occurs to some extent in *Luidia* and the astropectinids and more so in the goniasterids.

In contrast to the Ambb and Adambb, the marginals do not appear to show a more or less continuous sequence of change, although there is some suggestion of morphological continuity between the astropectinids and *Luidia*. Morphological diversity suggests that these ossicles have evolved much more rapidly than did the Ambb or Adambb.

Luidia Ossicle Morphology and the Classification of Döderlein (1920).—In his 1920 study, Ludwig Döderlein divided the genus *Luidia* into four "groups" and the four groups into ten subgenera. Relationships were suggested among these groups and subgenera. In his classification, Döderlein emphasized such external features as the form and appearance of primary ossicle systems and the development of spines, spinelets, and pedicellariae. In the present investigation, in order to evaluate the taxonomic value of skeletal elements within and above the species level, ossicles from various species of *Luidia* were compared to one another. In addition, the internal consistency of Döderlein's groups with respect to ossicle morphology was studied.

The species considered here represent all of Döderlein's four groups and eight of ten subgenera, and they are listed in table 1.

TABLE 1

SPECIES OF LUIDIA CONSIDERED IN THE PRESENT STUDY, AND THEIR CLASSIFICATION AFTER DÖDERLEIN, 1920

CLATHRATA GROUP	ALTERNATA GROUP
Subgenus *Senegaster*	Subgenus *Maculaster*
L. senegalensis	L. magnifica
Subgenus *Petalaster*	L. maculata
L. columbia[1]	L. mascarena
L. clathrata	Subgenus *Alternaster*
L. tesselata[2]	L. alternata
L. foliolata	L. phragma
CILIARIS GROUP	Subgenus *Armaster*
	L. ludwigi[4]
Subgenus *Hemicnemis*	**QUINARIA GROUP**
L. ciliaris	
L. asthenosoma	Subgenus *Quinaster*
L. elegans	L. quinaria
L. neozeolanica[3]	Subgenus *Penangaster*
	L. penangensis

[1] *L. brevispina* of Döderlein's usage; see A. M. Clark, 1953.
[2] *L. columbia* of Döderlein's usage; see A. M. Clark, 1953.
[3] Described since Döderlein's work, placed in this subgenus by Clark, 1953.
[4] Döderlein considered that this species might be a synonym of *L. armata*.

From the present study it appears that the InfMM of each group have a set of typical morphological features. Morphology of Ambb is less distinctive in each group, whereas Adamb morphology is the least distinctive of the three ossicle types. Because of insufficient available material, no comment can be made on the unity of the subgenera as reflected in ossicle morphology. The only species whose ossicle morphology suggests a group other than the one in which it was placed by Döderlein is *L. foliolata* Grube.

Fell (1963) elevated most of Döderlein's subgenera to generic rank. As discussed below, *Luidia,* in the sense of Döderlein, displays a general unity in ossicle morphology. The classification of Döderlein therefore is retained. Fell considered Döderlein's subgenus *Armaster* to be a synonym of another subgenus *Alternaster* but did not discuss his reasons. Döderlein's subgeneric classification is retained here, although study of ossicle morphology of further species might support placing some of the subgenera in synonymy.

Inferomarginal Morphology.—In *Luidia* the interspecifically most variable ossicles are the InfMM.

Because of the development of the paxillae and the small ossicles betwen the Adambb and the InfMM, *L. senegalensis* (Lamarck) was considered by Döderlein to be the most primitive of the extant luidiids. The InfMM of this species (pl. 2, figs. 11–22) have moderately well developed articulation ridges of angular outline. Articulation processes are distinct on medial and distal ossicles but are fused as an elongate ridge on ossicles taken from near the disc. The inner face step is gentle, the super ambulacral boss moderately prominent, and the oral side face not sharply inclined to the inner face. Three of the other four members of the Clathrata Group (in the sense of Döderlein) are similar in InfM morphology. *L. clathrata* (Say), suggested by Döderlein to be closely related to *L. senegalensis,* has InfMM (pl. 3, figs. 20–28) that are very similar to those of *L. senegalensis,* differing primarily in the development of the inner face of the ossicle. *L. columbia* (Gray) InfMM (pl. 1, figs. 13–24) are distinct from those of the other two species because of their outline and less massive development but possess similar articulation ridges and similar inner face development. *L. tesselata* (Lütken) InfMM (pl. 4, figs. 5–13), also less massive, are similar to those of *L. clathrata* and *L. senegalensis* in the development of the articulation ridges and the inner face step. The InfMM of the fourth species, *L. foliolata,* are similar to ossicles of Alternata Group species, rather than to ossicles of other Clathrata Group species.

L. magnifica Fisher was considered by Döderlein to be relatively closely related to primitive Alternata Group species because of the nature of paxillae development. *L. magnifica* InfMM (pl. 5, figs 26–31) have a wide, adradially tapering distal articulation ridge with a single prominent articulation process, a distinct inner face step, a prominent superambulacral boss, and prominent spine bases. *L. ludwigi* Fisher ossicles (pl. 5, figs. 10–15) have a distinct inner face step and superambulacral boss. The articulation ridges are similar in development to those of *L. magnifica* but bear elongate, partially disjunct processes. InfMM of *L. alternata* (Say), *L. maculata* Müller and Troschel, and *L. mascarena* Döderlein are all generally similar (pl. 6, 7). *L. phragma* Clark InfMM (pl. 8, figs. 10–15) have a distal articulation ridge similar to those of the above species, but with multiple articulation processes that, in ossicles of one individual observed, were disjunct. The inner face steps of the *L. phragma* ossicles are distinct, but not sharp. *L. foliolata,* considered to be a member of the Clathrata Group by Döderlein, has InfM morphology similar to that of *L. magnifica* of the Alternata Group. *L. foliolata* InfMM (pl. 9, figs. 12–20) have very prominent articulation ridges, each with a single articulation process, a sharp inner face step, and a prominent super-ambulacral boss.

Only *L. quinaria* v. Martens and *L. penangensis* de Loriol of the Quinaria Group are considered here. Their InfM morphology (pl. 7, figs. 14–19; pl. 8, figs, 22–27) is fairly distinct. The ossicles are proportionately small and massive, with moderately long continuous articulation ridges and an elongate articulation process. They are most similar to ossicles of *L. senegalensis* but are distinctive because of their massive form and the development of the articulation ridges.

The four Ciliaris Group members considered are very sharply set off from members of other groups. The InfMM (pls. 10, 11) are delicate and lack the steplike outline of other groups. The articulation ridges are long and wide. Various authors, including Mortensen (1925:280–281) and Fisher (1919:168) have expressed the opinion that many of the Ciliaris Group species are separated by rather minor differences. Differences in ossicle morphology between species are also minor.

First Paxillae Morphology.—Enlarged SupMM do not occur in *Luidia,* although the Fst Pax are probably homologous with the SupMM. Although the Fst Pax vary in size, morphological variation does not appear great; hence, these ossicles have not been intensively studied here.

Ambulacral Morphology.—Ambb reflect Döderlein's divisions fairly well and are the second most interspecifically variable arm ossicles within the Luidiidae. Ambb of members of the Quinaria and Ciliaris Groups are quite distinct from those of members of other groups. Those of the Quinaria Group (pls. 7, 8) are massive and symmetrical, with a low, rounded aboral ridge, except for a prominence at the adradial end of the aboral ridge. Those of the Ciliaris Group (pl. 10, 11) are delicate and relatively long, with a low, rounded aboral ridge. Ambb of species of the Clathrata Group (pls. 1–4) typically have more or less prominent, sharp aboral ridges and square or rectangular amb body outlines. On both sides of the amb body there are articulation surfaces consisting of a subcircular or elongate process and a depressed surface for a muscle. These structures, referred to under the abbreviation "N," are more or less equally developed, low, elongate, and symmetrical about the long axis of the ossicle. The oral apophyse is weakly developed in most species. Alternata Group Ambb (pls. 5–9) generally have more or less rounded aboral ridges, triangular amb body outlines, and unequally developed asymmetrical, high, subcircular interamb articulation structures (N). The oral apophyse is generally prominent. The differences in features between the two groups are not absolute. *L. tesselata* (pl. 4, figs. 21–26) Ambb have a prominent oral apophyse and interamb articulation structures (N) that are not symmetrical. *L. foliolata* (pl. 9, figs. 21–26) Ambb possess all the features of those of the Alternata Group, and these features are well developed in ossicles of this species. The amb body outline of *L. maculata* and *L. phragma* ossicles is somewhat rectangular.

Adambulacral Morphology.—Adambb are less variable in morphology than the InfMM or Ambb in the Luidiidae. Adambb of species of the Clathrata and Alternata Groups are quite similar; specific differences seemingly are more important than group differences, although development of certain features may be more typical of one group. For example, the intergroove depression (ps1) is typically higher adradially in Adambb of species of the Alternata Group. The ossicles of members of the Quinaria and Ciliaris Groups are distinctive, from each other and from those of members of the Alternata and Clathrata Groups. The Quinaria Group Adambb (pls. 7, 8) have prominent spine bases and a relatively small muscle depression on the distal face. The Adambb of the Ciliaris Group (pls. 10, 11) are relatively high, with features correspondingly narrow. The outer face is typically large and relatively flat.

The ossicle morphology, in general, of species of the Ciliaris Group is unified within itself but is distinctly set off from that of other groups. Consideration of further species of *Luidia* might support recognition of the Ciliaris Group as a separate genus.

Intraspecific Variation in Two Species of Luidia.—In order to obtain some measure of intraspecific variation, portions of a small number of specimens of *L. foliolata* and *L. clathrata* were disarticulated and studied. Other specimens were studied as much as possible without disarticulation.

The features of the InfMM, Ambb, and Adambb are consistent in development within both species. For example, the InfMM of *L. clathrata* are oval in outline, the inner face step gentle, and the aboral abradial corner notched for contact with the first paxillae. All proximal articulation ridges have prominent articulation processes that are disjunct, except in proximal ossicles of large specimens. Ambb are consistent in the development of their aboral ridge and Adambb, in their weakly developed adradial prominences (ad pm). Comparable variation exists in the *L. foliolata* specimens. Variation exists in both species, but not enough to lead to confusion in assignment.

Ontogenetic variation is greater than variation between conspecific individuals of similar size. In addition to ontogenetic variation between individuals of various ages and sizes, there is an ontogenetic variation in age and size of the various ossicles of an arm series. Variation of this second type is not strictly related to organism ontogeny because interbrachial marginals and proximal Ambb and Adambb of many species display variation related to their position and arm geometry.

In general, small InfMM are more massive than large InfMM. Distal ossicles of large specimens tend to be more massive than proximal ossicles of similar dimensions of smaller specimens. The inner face step in smaller ossicles may be a convex surface but is distinctly concave in larger ossicles. The number of articulation processes and spine bases is proportionately reduced on smaller ossicles. Distal ossicles and ossicles of immature specimens of different species and different groups frequently cannot be distinguished because defining structures are less prominent and less distinctly different. In larger ossicles, partial fusion of articulation processes may occur. Many of these features can be seen in the illustrations of the InfMM of *L. senegalensis* (pl. 2).

Interbrachial ossicles are relatively very wide, short, and cuneate in outline. Articulation ridges are reduced and similar (pl. 9, figs. 15–17). Although these changes are gradational, they are limited to a relatively small number of ossicles in a form with acute interbrachial angles, such as *Luidia*.

The small distal Adambb are proportionately narrower and higher than the large proximal Adambb of any given specimen. The surface features are correspondingly altered.

Ambulacral and adambulacral ossicles are more closely spaced near the oral region than farther out on the arm. This is probably an adaptation to increase the number of tube feet available in a given linear arm distance near the mouth; only a single tube foot is present between two sequential Ambb on one side of the arm of *Luidia*. Amb bodies of near-oral ossicles are relatively short and the aboral ridge in some species, relatively high and thin. The Adambb are compressed upon one another and therefore are more delicate in appearance. They are frequently more strongly overlapping and, as a result, are longer.

The Systematic Position of Luidia foliolata Grube.—Among the seventeen species of *Luidia* studied, only *L. foliolata* has ossicle morphology unlike other members of the group to which it was assigned by Döderlein. *L. foliolata*, assigned to the Clathrata Group by Döderlein, is similar to members of the Alternata Group in ossicle morphology (pl. 9). *L. foliolata* InfMM have very long distal articulation ridges, with a single, large articulation process, a prominent superambulacral boss, and a sharp inner face step. The side face outline and spine base development are similar to those of *L. alternata*. The Ambb have a distinctly triangular amb body, highly asymmetrical interamb articulation structures (N), and a low, rounded aboral ridge. Adamb morphology is most similar to that of *L. alternata*.

In discussing features that distinguish the groups, Döderlein observed that the Clathrata Group members have side paxillae that are very regular in size and orientation, that the medial paxillae are small, that the first paxillae are large, and that all paxillae lack spines. The aboral surface of the arm is flat. The Alternata Group members typically have small first paxillae, and, although the rows are regular, they are not densely spaced and therefore give the surface an irregular appearance. Very commonly, some paxillae bear aboral spines. Some of the many-armed Alternata Group members, especially *L. maculata*, have larger, more regularly arranged paxillae than the typical species of the Alternata Group. *L. foliolata*, with its large first paxillae and smaller side paxillae is similar to members of the Clathrata Group, but its morphology is not typical of that group in that the first paxillae are longer than they are wide (other members, in the sense of Döderlein, have first paxillae that are wider than they are long) and that the number of side paxillae exceeds the number of first paxillae (the numbers correspond in the Clathrata Group). In the Clathrata Group, large spines on the InfMM are generally restricted to the lateral margin; in the Alternata Group they extend onto the oral surface of the ossicle. In *L. foliolata* the spines extend onto the oral surface. The groove spines of the Adambb are more like those of Clathrata Group members. Pedicellariae are never present on members of the Clathrata Group, usually, but not always, occur on Alternata Group members, and are absent on *L. foliolata*. *L. foliolata* has irregular aboral coloring, a feature typical of the Alternate Group members, but absent from other species assigned to the Clathrata Group.

Because many features of the Clathrata Group used by Döderlein are atypical in their development in *L. foliolata* and because of the very strong similarities of the ossicle morphology of *L. foliolata* to that of the Alternata Group, *L. foliolata* is here assigned to the Alternata Group, probably to the subgenus *Armaster*, to which Döderlein assigned those species of the Alternata Group having five arms and normal or slightly reduced side paxillae.

The Value of Ossicles in Identification and Classification.—Species considered closely related by Döderlein are generally similar in ossicle morphology. InfMM, and possibly Ambb, of all specimens of all species seen, except possibly those of the Ciliaris Group, appear interspecifically distinct. It appears probable that the evolution of an individual ossicle type is more conservative than that of the species as a whole, because all of the *Luidia* species examined appear distinct if all three ossicle types are available.

Many of the variations in ossicle form with respect to position, discussed above for the Luidiiae, are also valid in other families. For example, in the Benthopectinidae, interbrachial marginals may be shorter, higher, and more symmetrical than those from the arm (compare pl. 16, figs. 1 and 2 with fig. 3). The arm Ambb possesses an aboral abradial process, which is used to brace the SupMM; this process is lacking in the Ambb of the disc (pl. 16, figs. 11 and 12). The Adambb of the disc are proportionately wider than those from the slender arms (pl. 16, figs. 13 and 14).

The importance of considerating multiple ossicle types is readily seen in the Astropectinidae (pl. 15). The gross morphology of *Thrissacanthias penicillatus* (Fisher) and *Dipsacaster eximius* Fisher, as well as marginal and Adambb morphology, are clearly distinct, yet the Ambb of the two species are quite similar. In contrast, it is the sharp, often spinose, aboral ridge of the Ambb that distinguishes *Astropecten ornatissimus* Fisher most readily (pl. 14, figs. 28, 29).

Knowledge of ossicle morphology of Recent starfish species is very limited; not enough is known to permit definition of species from isolated ossicles. Great care must be used in defining species from small fragments. Variation, however, appears to be constant and predictable; changes in InfMM interbrachially, or in near-oral Ambb or Adambb, are repetitive in nature from species to species. Isolated ossicles and fragments therefore show great promise as a taxonomic tool, not only to the paleontologist forced to contend with disarticulated material, but also to the neontologist seeking to define closely related species.

The Systematic Position of Platasterias latiradiata Gray.—Fell (1962, and subsequently) assigned the genus *Platasterias* to the subclass Somasteroidea, an interpretation that has been questioned (Madsen, 1966). On the basis of ossicle morphology, the writer believes *Platasterias* should be placed in the Family Luidiidae. The genus could be considered to be a distinct genus in the family, or, as Madsen suggested, it could be considered either as a member of the subgenus *Petalaster* of the Clathrata Group or as an aberrant subgenus of *Luidia*. *Platasterias*, because of its distinctive form, is here considered to be a distinct subgenus in the Clathrata Group. The essential features of *Luidia* ossicle morphology are present in *Platasterias;* only the proportions are different. The differences are significantly less than those that occur between species assigned to *Luidia* (compare *Platasterias*, pl. 1; *L. columbia*, pl. 1; *L. clathrata*, pl. 3; *L. penangensis*, pl. 5; *L. elegans*, pl. 11).

The first paxilla of *Luidia senegalensis* (pl. 2, fig. 3) is similar to those of *Platasterias* (pl. 1, figs. 30, 31), differing in certain relative proportions. The tabula and column of the *L. senegalensis* ossicle is relatively attenuated and reduced in width compared to the outer face and column of *Platasterias*. Exposed surfaces of both ossicle types are transversely elongate. Side face articulation structures are comparable in shape and position, but the base of the *L. senegalensis* ossicle is proportionately wider. In the presumably more advanced luidiids, the first paxillae are further reduced, becoming, in members of the Alternata Group, indistinguishable from other paxillae; in some taxa, Fst Pax are smaller than many paxillae from rows nearer the arm axis.

InfMM of *Platasterias* are most similar to those of species of the Clathrata Group. Articulation ridges of *Platasterias* InfMM are very short, with many articulation processes. The proximal ridge is formed by disjunct articulation processes. The inner face step of the *Platasterias* InfM is very gentle. The superambulacral boss is weakly developed. The *Platasterias* InfM differs most strikingly from the *Luidia* InfM in being relatively very wide and short. On distal ossicles these differences are less apparent. A distal *Platasterias* ossicle could be readily mistaken for the distal *Luidia* InfM (pl. 1, figs. 13–15; 44, 45).

The Amb of *Platasterias* is also very similar to that of *Luidia*, differing in that the aboral ridge is truncate adradially and, like the InfM, in relative width. The Adamb differs only in relative width. In both ossicle types, distal ossicles differ less in width. Therefore, *Platasterias* ossicles of all four types, especially the InfM, Amb, and Adamb, are very similar to those of the Clathrata Group. The ossicles, like the entire animal, are relatively wide and flat, possibly, as suggested by Madsen, an adaptation to a shifting sandy bottom or to feeding habit.

Ossicle morphology of *Platasterias* appears relatively distinct from that of the Ordovician somasteroids, insofar as can be inferred from illustrations and descriptions of these genera (Spencer, 1951; Spencer and Wright, 1966). For example, marginals of the fossils are elongate parallel to the arm. They also are massive and appear relatively simple in morphology, apparently lacking the complex articulation structures and spine base pattern found in *Platasterias*. Both Ambb and Adambb of the fossils appear relatively long and massive when compared to those of *Platasterias*. Other structures, such as basins for the tube feet and the adradial tongue-shaped projections of *Archegonaster*, are found on the Ambb of the fossils, but not on those of *Platasterias*.

Ossicle morphology does not support the somasteroid position suggested for *Platasterias*, but rather supports the inclusion of *Platasterias* within *Luidia* (Blake, 1972).

Comparative Skeletal Features in Asteroids.—Marginal ossicle morphology in the Asteroidea is very diverse. The marginals of the Luidiidae and certain members of the Astropectinidae are morphologically complex and very similar, suggesting that they are homologous. The present sample is too small and the marginal morphology too diverse to recognize possible homologies in other taxa, beyond the broad morphological similarity seen within the Benthopectinidae or Goniasteridae. The simplicity of form and general lack of well differentiated articulation structures in the marginals of families such as the Goniasteridae suggest that these ossicles may be of limited use in the recognition of broad evolutionary paterns.

The Adambb and Ambb display more limited, but distinct, variability. These ossicles are linked to one another by a complex arrangement of muscles and articulation processes. Although greatly varied in proportions, certain analogous and probably homologous structures occur throughout the small, but taxonomically diverse, sample considered here. Information derived from these ossicle structures and relationships appears useful in evolutionary interpretations.

Aboral ossicles such as the paxillae are limited in variation but do appear to warrant further comparative study.

Comparisons Between Members of the Luidiidae and the Astropectinidae.— SupMM of *Platasterias* (pl. 1, figs. 30, 31) are similar to those of the astropectinid, *Dipsacaster eximius* (pl. 15, figs. 9–11). In both species, the InfMM are considerably larger than the SupMM. The SupM possesses an adradially projecting process that articulates with the neighboring Pax and InfM. In both species, the outer face is raised on a pedestal-like portion of the ossicle, and side face articulation structures are similar. Some of these features also occur in the SupMM of *Astropecten regalis* Gray (pl. 14, figs. 16, 17), but in this species the features are less prominently developed.

All three species are flattened and have relatively broad arms. The arms of *A. regalis* and *Platasterias* are petaloid. The outline of the arm of *D. eximius* is curved; taper increases distally, creating a subpetaloid appearance. The ossicle similarities are probably in part a product of similar overall form, which in turn may be a product of adaptation to a similar ecological habitat.

The InfMM of members of both families are basically similar but differ prominently in outline and articulation ridge development. *Luidia* InfMM have a more or less distinct inner face step (text fig. 2B). The articulation ridges are wide, with distinct notches in the ossicle for the articulation ridge overlap. The ridges also bear multiple or elongate articulation processes. The articulation processes are only slightly raised above the surface of the ossicle. *Astropecten* InfMM are massive; the side face outline is ovate or triangular. Overlapping articulation ridges are absent, and the articulation processes are partially fused into a continuous ridge.

Many features are similar between the genera; both are transversely elongate, with ovate (or "notched" ovate, in the case of certain luidiids) outlines, both possess polygonal inter-InfM articulation structures in their adradial aboral quadrants, and both have SupM (or first paxilla, in *Luidia*) articulation surfaces abradially on the aboral surface. In both, fasciolar surfaces occupy the abradial region and the oral margin of the side faces, and many spine bases occupy the short, wide abradial and oral surfaces.

InfMM of *Dipsacaster eximius* are of this same pattern, but those of *Thrissacanthias penicillatus* and *Psilaster pectinatus* (Fisher) are relatively very narrow, and the articulation areas are high.

The first paxillae of *Luidia* here are considered to be homologous with the SupMM of the Astropectinidae. In a descriptive sense, the differences between Fst Pax of Clathrata Group members and SupMM of the astropectinids are differences of size and proportion, not of basic shape. The simplest relationship between these two ossicle types would be to derive one from the other, either by an increase or a decrease in size, with associated changes in proportion. If, following Döderlein, *L. senegalensis* presents a reasonable approximation of a primitive *Luidia,* then both processes may have taken place within the Luidiidae—reduction, toward the condition seen in the Alternata Group, and inflation, as seen in the development of the Fst Pax of *Platasterias.*

Although the Fst Pax are considered to be homologous with the SupMM, both terms are retained because the differentiation is descriptively useful. The Fst Pax do appear to be more a part of the paxillae field in *Luidia;* the inflated SupMM, a part of the marginal series in the Astropectinidae.

InfMM of both families are also morphologically very similar; the simplest explanation for these similarities is, again, homology. If the InfMM are accepted as homologous, then it becomes even more probable that the SupMM and first paxillae are homologous as well, because in this situation the SupMM and first paxillae would both be in the same position relative to ossicles of accepted homology.

The mouth ossicles of *Platasterias, Luidia,* and *Astropecten* are similar (text figs. 2J–2O) in overall form, spine base development, and aboral articulation structures. *Platasterias* mouth ossicles are relatively wide. Ossicle proportions and the enlarged distal intermouth ossicle articulation structures in *Platasterias* suggest that the greater relative width may be primarily on the distal part of the ossicle.

Comparative Morphology of Ambulacral Ossicles.—Ambb of all considered taxa are similar in being cylindrical structures oriented normally to the arm axis. They articulate adradially with the opposite member of the Amb pair. Articulation is by means of the dentition and muscle that attaches to the oral muscle groove (Unt G) and another depression, not considered elsewhere here, located aboral to the dentition on the adradial face. Longitudinally, they articulate with adjacent Ambb by means of the interamb articulation structures (N) and abradially with the Adambb by means of the lateral wings (GG), and the adamb notch. As far as can be determined from the present sample, the form of the amb body and Adamb articulation appear to have undergone consistent patterns of change in evolution.

In the Luidiidae and Astropectinidae, the adamb notch is an arced surface whose long axis is essentially parallel to the axis of the ossicle. In the Goniasteridae, the articulation surface is inclined to the axis of the ossicle (pl. 18, figs. 6, 16, 29). Because this surface is approximately parallel to the oral surface in the living animal, the change in orientation may have been an adaptation to provide the asteroid with a deeper, narrower ambulacral furrow. In addition, species in the Luidiidae and Asteropectinidae have only a single Adamb articulation apophyse, located adradially from the adamb notch. The goniasterids have articulation apophyses on both ends of the articulation surface.

The Amb of *Astropecten* and the Clathrata Group are similar in being symmetrical; they have rectangular amb bodies, symmetrical interamb articulation structures (N), and narrow, keel-like aboral ridges. Ambb of the Alternata Group, other astropectinids, and goniasterids have overlapping, and therefore asymmetrical, amb bodies.

The proximal part of the amb body overlaps the distal margin of the adjacent

Amb. Overlap is more pronounced in the goniasterids than in the astropectinids or *Luidia*. Overlap may have been an adaptation to reduce Amb spacing, thereby increasing the number of Amb pairs in a given linear distance of arm. Overlap also provides a more extensive articulation area.

The amb bodies of the Alternata Group, other astropectinids, and the goniasterids are triangular; in these species, the aboral ridges are rounded. The sinuous outline of the Amb of *Ceramaster leptoceramus* (Fisher) (pl. 18, figs. 4, 5, 7) may be a result of the need to provide passage for the tube foot, yet retain the extensive articulation surfaces between neighboring Ambb.

Some convergent evolution in ossicle form appears associated with convergence in arm form; for example, the amb bodies of both the benthopectinid *Cheiraster gazellae* Stüder and the goniasterid *Paragonaster ctenipes hypacanthus* Fisher are relatively long and rectangular.

Comparative Morphology of Adambulacral Ossicles.—The morphological features of the Adambb of *Thrissacanthias* and *Dipsacaster* (pl. 15, figs. 1, 2, 25–27) are very similar to those of *Luidia*. Those of *Astropecten* are relatively distinctive (text figs. 2A–2F). Nevertheless, the ossicles of *Astropecten* have the same orientation in the living animal and bear all the essential features of those of *Luidia*. The ambulacral ossicle articulation groove (abbreviated pa1) of *Astropecten* is very similar to that of *Luidia*. In *Astropecten,* the groove (pm1) for the muscle that extends to the proximal lateral wing (pGG) of the abradial part of the Amb, termed *ambulacral extension,* or *amb extension,* is a small adradial subtriangular or trapezoidal feature, rather than a relatively wide structure, as in *Luidia*. The groove (pm2) for the muscle that extends to the distal lateral wing (dGG) of the ambulacral extension is a very large structure occupying nearly all of the area between the horizontal interadambulacral articulation surface, termed the *pa4,* and the pa1 and pm1. In *Luidia*, the pm2 is wide, but relatively low; therefore, the intergroove depression (psr1) is relatively large. In *Astropecten,* the psr1 is a very small, abradial, triangular surface immediately oral to the pa1. In both genera, the pa4 is very prominent and elongate and in general appearance divides the surface into two areas. However, the structure in *Astropecten* is horizontal (relative to the ossicle surface on which it lies) and parallel-sided. In *Luidia*, it is inclined and cuneate or ovate. The vertical interplate articulation process (pa3), which provides a contact between successive ossicles, is similar in form and position in both genera. The process that the pa4 contacts on the distal face of the next proximal Adamb is a weakly developed surface (da1) near the aboral margin. The pa3 contacts a medial adradial process (da2) on the next Adamb. In both genera, these four features are similar in position and development. The process that articulates with the oral apophyse of the Amb (pa2) has a similar position in both genera. The interadambulacral articulation depression on the proximal face (pm3) is in *Astropecten,* as in *Luidia,* essentially a triangular area occupying the remaining space near the oral margin. The interadambulacral articulation depression on the distal face (dm1) is also very similar in form in these genera.

A considerable change of orientation, and therefore structure, has taken place in the Adambb of the Benthopectinidae. Fewer structures are distinctly developed.

The most significant difference is in the nature of the Amb-Adamb articulation. The prominent articulation keel present along the abradial midline of the oral surface of the Amb of *Luidia* and *Astropecten* is, in the benthopectinids, a weakly developed, rounded ridge offset proximally from the ossicle midline. Most of the abradial oral surface is essentially flat and continuous with the oral surface of the distal lateral wing (dGG) (pl. 16, figs. 9, 10, 28, 29). Muscle extends from this area to a much-enlarged pm2 of the Adamb (text fig. 2M). Muscle from the small proximal lateral wing (pGG) extends to a usually inconspicuous pm1 along the aboral margin of the proximal face of the adamb. The pm1 and pm2 form essentially a continuous surface. The ambulacral ossicle articulation groove (pa1) is an inconspicuous process immediately adradial to the pm1. The Amb is further supported by the Adamb at an inconspicuous contact along the abradial margin of the proximal face and by the pa2, which is continuous with the aboral of the interadambulacral articulation processes, the pa3 and pa4. The proximal face interadambulacral muscle depression (pm3) and the adradial prominence (ad pm) are well developed. The distal face interadambulacral muscle depression (dm1) and one of the distal face interadambulacral articulation processes (da2) are well developed, whereas a second process (da1) is weakly developed.

Fisher (1911:20) and Verrill (1899:198–200) both believed *Archaster typicus* Müller and Troschel to be a genus of uncertain relationships and considered it best separated from other genera by placing it by itself in the Archasteridae, as suggested by Verrill.

Although displaying the same basic features and distribution of features seen in *Luidia* and the Astropectinidae, *Archaster* Adambb possess structures not seen on Adambb of other taxa (text figs. 2J–L).

As in Adambb of *Astropecten* and the Benthopectinidae, the Amb articulation structures are more distinct from the Luidiidae pattern than are the inter-Adamb structures, suggesting that these latter features are evolutionarily more conservative.

Archaster Adambb possess a moderately prominent adradial prominence (ad pm). The pattern of the dm1, da2, pm3, and pa3 are similar to that seen in *Luidia*. The ambulacral ossicle support process (pa2) is the abradial portion of a prominent process whose adradial side contacts a small ossicle that fits between the Adamb and Amb. This ossicle is termed an *exambulacral* by Schäfer (1962). One of the depressions for muscle linking the Adamb with the lateral wing (GG) of the Amb, the pm2, is very weakly developed. The other such depression, the pm1, tapers adradially, rather than abradially as in *Luidia*, but both features are in positions comparable to that found in *Luidia*. The ambulacral ossicle articulation groove (pa1) is hook-shaped, as in *Luidia*, but is a two-part, positively arched structure, rather than the negative continuous groove found in *Luidia*. The interadambulacral processes, da1 and pa4, appear to be lacking, or extremely weakly developed.

Archaster Adambb are distinctive in the possession of a prominent proximal-face articulation process, which contacts the abradial margin of the oral surface of the distal lateral wing (dGG) of the Amb (text figs. 2J, 2K; abbrev., pa5), a prominent abradial process that contacts the neighboring InfM (text figs. 2J, 2L;

abbrev.: da3) and a prominent groove for a muscle that extends between the Adamb and InfM (text figs. 2L; abbrev.: dm2). These are all structures that would appear to provide a firmer linkage among the various ossicles around the ambulacral furrow.

The Adambb of some goniasterids (pl. 18, figs. 1–3, 26–28) are much more regular in shape, with only a slight distal asymmetry (text fig. 3). A distinct aboral surface is developed. The surface has six articulation areas, an adradial and abradial one along the proximal and distal margins and two elongate ones extending transversely across the ossicle between the two pairs of marginal processes. The distal aboral abradial process forms a V with the proximal aboral abradial process of the next distal Adamb; the two adradial articulation processes likewise form a V, and it is against these Vs that the base of the Amb rests. A muscle from the proximal lateral wing (pGG) is attached to the elongate distal articulation groove of the Adamb proximal to the Amb. Muscle from the distal lateral wing (dGG) is attached to the elongate proximal groove of the Adamb offset distal to the Amb.

The aboral surface is considered comparable to that portion of the *Luidia* ossicle aboral to the pa4 and extending to the pa1. The aboral distal articulation processes are functionally equivalent to the pa1 and the adradial aboral proximal process

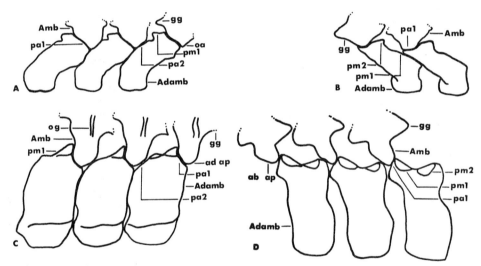

Fig. 3. Relationships between Ambb and Adambb in *Luidia foliolata* Grube and *Protoreaster nodosus* (Linné).

A, *L. foliolata,* adradial view, proximal left.

B, *L. foliolata,* abradial view, proximal right, both hypotype no. 10685; × approx. 6½.

C, *P. nodosus,* adradial view, proximal left.

D, *P. nodosus,* abradial view, proximal right, both hypotype no. 10686; × approx. 6½. The Adambb of *Luidia* arch strongly distally; those of *Protoreaster* have only a small distal inclination.

The Amb of both species is supported by two Adambb adradially; but the Amb of *L. foliolata* is supported only by the proximal Adamb abradially, whereas the Amb of *P. nodosus* is supported by both Adambb. *Luidia* has no abradial structure equivalent to the pa2, but an equivalent structure is formed abradially in *P. nodosus.*

Members of the Luidiidae and Astropectinidae considered possess the basic *L. foliolata* pattern; those of the Oreasteridae and Goniasteridae have the *P. nodosus* pattern.

equivalent to the pa2. The elongate muscle depressions between the paired processes are equivalent to the pm1 (the distal depression) and the pm2 (the proximal depression).

In *Luidia,* the Amb contacts two sequential Adambb adradially; the pa1 of the proximal Adamb and the relatively inconspicuous pa2 of the distal Adamb. Abradially, the Amb is restricted in contact to the pa1. There is no abradial structure comparable to the pa2. Muscles link the two ossicles abradially, between the pm2 of the Adamb and the distal lateral wing (dGG) of the Amb. In *Mediaster,* articulation is nearly symmetrical, as are the articulation structures. The Amb apparently has been dislocated distally in the course of evolution, so that an Amb of *Mediaster* is between the Adambb, rather than resting primarily on a single Adamb (text fig. 3). As discussed above, correlated with the development of the adradial and abradial supports on the Adamb is the development of an adradial and abradial apophyse on the Amb.

The proximal and distal faces of the Adamb of *Mediaster* are dominated by the depressions for the muscles that link successive Adambb. These are interpreted as the pm3 and dm1. Generally, weakly developed inter-Adamb contact surfaces are present along the adradial margins of the proximal and distal faces. These surfaces are in the correct position to be equivalent to the da2 and pa3 of *Luidia.*

Some goniasterids with long slender arms, such as *Paragonaster,* have relatively long, narrow Adambb with less distinctly developed articulation structures (pl. 18, figs. 17, 18) than those found in the broader-armed species.

Adambb of *Nidorellia armata* (Gray), of the Family Oreasteridae, are very similar to those of *Mediaster* (text figs. 2Q, R). There are paired aboral articulation processes along the margins of the aboral surface and two elongate, well developed articulation surfaces between these processes for the muscles extending to the lateral wings (GG) of the Ambb. The side faces are dominated by the inter-Adamb muscle depressions and are bordered by weakly developed, low interplate articulation surfaces.

These structures are interpreted as homologous with corresponding structures in *Mediaster.*

Dermasterias imbricata (Grube) and *Patiria miniata* (Brandt) of the Order Spinulosida both have short, triangular Adambb that are crescentic in outline when viewed from the ambulacral furrow.

Morphological features are somewhat inconspicuous but are comparable in relative position and nature to those of Adambb of genera of other families. The Amb articulation processes and side face inter-Adamb muscle depressions are best developed (text figs. 2U–X).

These structures are likewise considered homologous with those of taxa already described.

Asteroid Form Reflected in Ossicle Morphology.—In some instances it is possible to reconstruct a part of the overall starfish form from a few ossicles. Such information can be of use to the paleontologist who has available only a few fossil fragments.

The cuneate interbrachial marginals of many species, for example, *Astropecten armatus,* suggests acute interbrachial angles. The same species has relatively high marginals, especially the SupMM. The arm margin in the living animal is also high and massive in nature. The development of prominent articulation ridges and small spinelet bases on the side faces of marginals suggests the development of fasciolar grooves. Spine development is suggested by the size of spine bases and their density on the outer face. The presence of small, superambulacral bosses on InfM ossicles suggests narrow arms because the InfMM had to be close to the Amb to be used for a superambulacral support. On the InfMM of *Dipsacaster,* the intermarginal articulation processes are well back from the ossicle margin, suggesting that the InfMM alone formed the arm margin. In the Benthopectinidae, the Ambb have a very high aboral ridge; the ridge has a well-developed abradial prominence, used to brace the SupMM. The arms, therefore, are only slightly wider than an Amb pair, and the necessity for a brace suggests the fragile nature of the arms.

West American Fossil Starfish.—Fragmentary or complete fossils are very rare. The few described here all belong to families that now occur in the northeast Pacific. With the exception of the Cretaceous *Sucia suavis,* n. gen., n. sp., all belong to living genera or to genera that appear closely related to living genera. There is no evidence for any significant change in the nature of the fauna during the Cenozoic.

SYSTEMATIC DESCRIPTIONS

Descriptions of all taxa are based primarily on ossicles taken from medial and proximal sections of arms of mature specimens. Only features found to be useful in taxon discrimination have been included; many morphological features are undescribed. A key to the groups of *Luidia* is included here, but keys to other taxa are not, because the species sample, selected to provide diversity, is not comprehensive for any other taxonomic category or geographic area; an identification made from any key based on such a sample could be misleading. Features found most distinctive for each species are summarized in the descriptions.

Abbreviations used are explained in the glossary.

Subphylum ASTEROZOA

Class STELLEROIDEA

Subclass ASTEROIDEA

Order PLATYASTERIDA

Family LUIDIIDAE

Genus *Luidia*

KEY TO THE GROUPS OF LUIDIA

1. IM outline crescentic, distal articulation ridge wide, ridge outline rectangular; Amb low, more or less fragile; Adamb proportionately narrow with narrow, triangular pml and large psrl ....Ciliaris Group

    IM outline cuneate, distal articulation ridge outline more or less triangular; Amb relatively massive; Adamb proportionately wide with wide pml and more or less small psrl 2

2. IM massive, distinct inner face step lacking, articulation processes on ridges elongate, linear; Amb massive, amb body symmetrical, subrectangular, aboral ridge broad, rounded, proximal part of ridge inflated; Adamb outer face large ....Quinaria Group

    IM fragile-to-moderately-massive, articulation processes on ridge disjunct, or a single elliptical process; Amb fragile-to-moderately-massive, aboral ridge not broadly rounded, amb body rectangular or triangular; Adamb outer face small 3

3. IM fragile-to-moderately-massive, inner face step more or less gentle, outer face outline parallel-sided, distal articulation ridge short, outline concave-to-gently-convex, articulation processes typically disjunct, although in some species there may be some fusion of adradial processes; amb body rectangular, aboral ridge distinct, relatively sharp ....Clathrata Group

IM relatively fragile, inner face step distinct, with a well-defined superambulacral boss, outer face outline more or less irregular, generally with well-defined spine bases; distal articulation ridges wide, outline convex, usually with a single elliptical articulation process, although in some species there may be some tendency toward disjunct adradial processes; Amb body outline triangular, aboral ridge low, rounded .... Alternata Group

Species of the Clathrata Group of Döderlein, 1920

The following features occur in ossicles of Clathrata Group species, except as modified under individual descriptions.

Inferomarginal ossicles: Inner face step small-to-moderately-sharp, inner face surface gentle; oral side face inclined to inner face; outer face long, parallel-sided; distal articulation ridge short, outline concave-to-gently-convex; proximal articulation ridge corner perpendicular, vertical component linear; multiple articulation processes, commonly some fusion of processes, usually the adradial processes.

Ambulacral ossicles: Amb body symmetrical, rectangular, GG asymmetrical, margins parallel, pGG tapers abradially, dGG tapers adradially, pGG separated from amb extension by sharp groove; Adamb notch curved, N similar, aboral ridge sharp, prominent adradially.

Luidia (Platasterias) latiradiata (Gray), 1871
(pl. 1, figs. 28–53)

Platasterias latiradiata Gray, 1871:136–137, pl. 9, figs. 1, 2; Fell, 1962:1–16, pl. 1–4, text figs. 1–8.

Description.—InfM (pl. 1, figs. 44–53). Ossicle low, somewhat massive, outline angular, elongate, ovate; inner face step very gentle, superambulacral boss weak, located near adradial margin; oral side face inclined to inner face; abradial region of ossicle low, triangular; oral abradial corner angular; outer face parallel-sided, sinuous, without abradial constriction; face with transverse row of about six widely spaced moderate-sized spine bases; distal articulation ridge very short, outline concave, adradial junction notched, abradial junction perpendicular; proximal articulation ridge corner perpendicular, vertical component of ridge linear; many separate articulation processes. InfM comparisons: Distinguished by short, wide proportions.

Amb (pl. 1, figs. 38–43). Amb body outline rectangular, dentition primarily of transverse plates; oral apophyse weakly developed, GG asymmetrical, margins parallel, pGG tapers abradially, dGG tapers adradially, pGG separated from amb extension by sharp groove, Adamb notch linear; N not prominent, similar, elliptical; aboral ridge sharp, truncate adradially. Amb comparisons: Distinguished by relative proportions, elongate Adamb articulation surface.

Adamb (pl. 1, figs. 32–37). Short, very wide ossicle, outline rectangular; ad pm prominent, pm1 very wide, psr1 low, tapers adradially, pm3 low, elongate; dm1 shallow; spine bases weak. Adamb comparisons: Distinguished by relative proportions; distal Adambb are relatively narrower than proximal ones and may be indistinguishable from those of certain other species.

Fst Pax (pl. 1, figs. 30, 31). Ossicle massive, tabula quadrate, column short, articulation projections weakly developed; much larger than neighboring Pax.

Pax (pl. 1, figs. 28, 29). Ossicle small; tabula subquadrate; column short, articulation projections fairly prominent; medial ossicle tabula subcircular, column very short, base irregular.

Material.—Two specimens, UCMP B-8740; three specimens, USNM E6612 and E8365; all from near Corinto, Nicaragua; Recent; ossicle descriptions based on proximal and medial ossicles of one UCMP specimen of R = 64 mm, r = 11 mm; hypotype no. 10634.

Luidia senegalensis (Lamarck), 1816
(pl. 2, figs. 1–31)

Asterias senegalensis Lamarck, 1816 (vol. 2) :567.
Luidia senegalensis: Döderlein, 1920:249–250, pl. 18, fig. 9; pl. 20, fig. 20.

Description.—InfM (pl. 2, figs. 11–22). Ossicle moderately high, massive, outline ovate; inner face step small, superambulacral boss moderately prominent; abradial region elliptical, long axis approximately vertical, oral surface outline gently convex, oral abradial corner gently rounded; outer face long, parallel-sided with slight abradial constriction, with transverse moderate-sized spine base row along abradial oral margin, a few small spine bases, especially along ossicle margins; distal articulation ridge outline linear or gently concave, adradial junction notched, abradial junction inclined, proximal articulation ridge corner slightly obtuse; multiple articulation processes, some adradial fusion of processes forming a linear ridge. InfM comparisons: Most similar to InfMM of *L. clathrata;* distinguished by the inclined abradial junction, the rounded outline of the abradial region and the relatively larger spine bases.

Amb (pl. 2, figs. 23–31). Dentition irregular, consisting of transverse and longitudinal plates; oral apophyse weakly developed; N similar, moderately prominent, elongate, elliptical; aboral ridge high, moderately sharp; adradial part of ridge prominent, aboral ridge of proximal ossicles very high. Amb comparisons: Distinguished by square amb body, high aboral ridge, and weak oral apophyse.

Adamb (pl. 2, figs. 4–10). Ossicle rectangular, wide; ad pm prominent, angular; pm1 wide, psr1 low, pa3, pa4 not prominent; pm3 triangular, high, with oral flange; spine bases moderate, not discrete. Adamb comparisons: Distinguished by the very low psr1, high flanged triangular pm3, and wide pm1.

Fst Pax (pl. 2, fig. 3). Ossicle moderately massive; tabula quadrate, rectangular; column short, massive, base massive, with prominent articulation projections.

Pax (pl. 2, figs. 1, 2). Ossicle relatively short, massive, tabula rounded on medial ossicle, quadrate on lateral Pax; column short, slender, base massive, with prominent articulation projections.

Material.—One unnumbered specimen, HMS, Florida; numerous specimens, USNM, including E16600, E6765, Florida, Trinidad, and Brazil; all Recent; ossicle descriptions based on arm of the HMS specimen of R = 105 mm to 130 mm, r = 23 mm, hypotype UCMP no. 10635.

Luidia clathrata (Say), 1825
(pl. 3, figs. 1–28)

Asterias clathrata Say, 1825: 142.
Luidia clathrata (Say) Lütken: Verrill, 1915:200–201, pl. 24, fig. 2.

Description.—InfM (pl. 3, figs. 20–28). Ossicle high, massive, outline ovate; inner face step gentle, superambulacral boss moderately prominent, oral side face gently inclined to inner face; abradial region squared, high; oral surface outline gently convex, oral abradial corner gently rounded, outer face parallel-sided, not constricted abradially, transverse row of moderate-sized spine bases along outer face, numerous small-to-moderate spine bases moderately densely arranged around larger bases, size differentiation small; distal articulation ridge outline gently convex, adradial junction notched, abradial junction perpendicular, with multiple articulation processes. InfM comparisons: Distinguished by relatively massive form and small spine bases.

Amb (pl. 3, figs. 11–19). Dentition consists primarily of longitudinal plates, weakly developed medially; oral apophyse moderately developed; N moderately prominent, similar in appearance, elongate, elliptical; aboral ridge sharp, very prominent, very sharp adradially. Amb comparisons; Distinguished by the high, sharp aboral ridge.

Adamb (pl. 3, figs. 5–10). Ossicle rectangular, high, ad pm small; psr1 moderate, pm3 triangular, high; spine bases prominent. Adamb comparisons: Distinguished by rounded ad pm, prominent spine bases, and high, triangular pm3.

Pax (pl. 3, figs. 1, 3). Column massive, short; tabula aproximately square on marginal ossicle, circular on medial ossicle, with densely arranged spinelet bases; column massive with numerous strong articulation projections; Fst Pax (pl. 3, figs. 2, 4) similar to Pax, larger, with rectangular tabula, articulation projections not prominent.

Material.—About 17 specimens in UCMP; numerous USNM specimens, including E5229, 1976,

3362; all Recent; Gulf of Mexico, Caribbean, Atlantic coast of the United States; ossicle descriptions based primarily on arm of one UCMP specimen of R = 90 mm, r = 14 mm, UCMP A-7557; hypotype no. 10636.

Luidia columbia (Gray), 1840
(pl. 1, figs. 1–27)

Petalaster columbia Gray, 1840:183.
Luidia brevispina: Döderlein, 1920:253–254, pl. 18, fig. 10; pl. 19, fig. 14; pl. 20, fig. 22.
Luidia columbia: Clark, A. M., 1953:381–383, text figs. 1, 2; pl. 39, fig. 1.

Description.—InfM (pl. 1, figs. 13–24). Ossicle low, outline subelliptical, elongate; inner face step moderate, superambulacral boss moderately prominent; abradial region elliptical, long axis horizontal, oral surface outline gently convex, oral abradial corner broadly rounded; outer face with abradial constriction; with numerous small-to-moderate-sized, moderately densely arranged spine bases, with smaller bases near proximal margin; distal articulation ridge outline gently convex, adradial junction notched, abradial junction inclined; proximal articulation ridge corner obtuse, vertical component curved; multiple articulation processes, some fusion of processes adradially. InfM comparisons: Distinguished by low, elliptical outline, notched aboral abradial margin.

Amb (pl. 1, figs. 7–12, 25–27). Dentition consists of transverse plates, with medial gap; oral apophyse moderately developed; pN very prominent, forms ridge, outline subcircular, dN slightly inset into aboral ridge; aboral ridge moderately sharp, with medial prominence; adradial part small, distinct from N. Amb comparisons: Distinguished by square amb body, prominent subcircular N and a medial prominence on the aboral ridge.

Adamb (pl. 1, figs. 1–6). Ossicle wide, rectangular, ad pm prominent, angular; psr1 low, triangular, with oral flange; numerous discrete, small adradial spine bases. Adamb comparisons: Distinguished by low psr1, low, flanged, triangular pm3, prominent angular ad pm, and discrete spine bases.

Pax. Ossicle low, tabula square on marginal ossicle, circular on medial ossicle, with small, densely arranged granule depressions; column massive, low, base massive with prominent articulation projections; Fst Pax similar, with rectangular tabula, very prominent abradial articulation projections.

Material.—Two specimens, UCMP B3649, Galapagos Islands; two specimens, USNM 39893, 40011, Pacific coast of Mexico; all Recent; ossicle description based on UCMP specimen of R = 67 mm, r = 13 mm; hypotype no. 10637.

Luidia tesselata Lütken, 1859
(pl. 4, figs. 1–26)

Luidia tesselata Lütken, 1859:40–42.
Luidia columbia Döderlein, 1920:253.
Luidia tesselata Clark, A. M., 1953:382–383.

Description.—InfM (pl. 4, figs. 5–13). Ossicle moderately high, outline cuneate; inner face step moderate, superambulacral boss placed adradially, moderately prominent; abradial region high, squared; oral surface outline linear, oral abradial corner gently rounded; outer face with very slight abradial constriction, with two large, abradial spine bases, transverse row of approximately five moderate-sized bases on oral surface, small-to-moderate spine bases loosely spaced on remainder of surface, concentrated near proximal margin; distal articulation ridge outline concave-to-linear, adradial junction notched, abradial corner perpendicular; proximal articulation ridge corner obtuse, with multiple articulation processes, some fusion of processes adradially. InfM comparisons: Distinguished by cuneate, squared outline, moderate spine bases, adradial superambulacral boss, and concave or linear distal articulation ridge outline.

Amb (pl. 4, figs. 21–26). Amb body subrectangular; dentition transverse, with medial gap; oral apophyse prominent; N moderately prominent, elongate, elliptical, pN prominent, dN slightly inset; aboral ridge sharp, with well-developed adradial prominence distinct from N. Amb comparisons: Distinguished by subrectangular amb body, elongate elliptical N.

Adamb (pl. 4, fig. 14–20). Ossicle wide, rectangular; ad pm moderately prominent, rounded;

psrl low, pm3 subquadrate with small oral flange. Adamb comparisons: Distinguished by low psrl and low, subquadrate, flanged pm3.

Fst Pax (pl. 4, figs. 3, 4). Ossicle relatively wide, high; tabula rectangular, with small, densely spaced granule depressions; tabula grades into column, forming triangular structure; base with massive, not strongly prominent, articulation projections.

Pax (pl. 4, figs. 1, 2). Generally similar to Fst Pax; ossicle near margin with square tabula, tabula of medial ossicle circular; column moderately high, light, with distinct articulation projections.

Material.—Two specimens, UCMP B-5854; two specimens, USNM 51115, 51116; all from Baja California; Recent; ossicle descriptions based primarily on arm of UCMP specimen of $R = 200$ mm, $r = 18$ mm; hypotype no. 10638.

Species of the Alternata Group of Döderlein, 1920

The following features occur in ossicles of Alternata Group species, except as modified in individual descriptions.

Inferomarginal ossicles: Inner face step deep, sharp, moderately prominent or prominent superambulacral boss; oral side face inclined to inner face, outline of outer face irregular, with prominent spine bases; distal articulation ridge outline convex, corner of proximal ridge obtuse, vertical component curved, with one large, approximately equidimensional articulation process.

Ambulacral ossicles: Amb body outline triangular; GG asymmetrical, margins parallel, pGG tapers abradially, dGG tapers adradially, junction between pGG and amb extension smooth, aboral ridge low, rounded, adradial part low, rounded.

Luidia magnifica Fisher, 1906
(pl. 5, figs. 18–34)

Luidia magnifica Fisher, 1906:1033–1036, pl. 15, figs. 1–3; pl. 16, fig. 1.

Description.—InfM (pl. 5, figs. 26–31). Ossicle moderately high, outline subelliptical; superambulacral boss prominent; oral side face continuous with inner face; abradial region elliptical, long axis horizontal; oral outline convex, oral abradial corner broadly rounded; outer face with slight abradial constriction, with transverse row of approximately five very large spine bases on outer face, numerous small-to-moderate-sized spine bases loosely distributed on remainder of outer face; distal articulation ridge length moderate, outline convex, adradial junction notched, abradial junction inclined. InfM comparisons: Distinguished by elongate abradial region, large spine bases, and oral side face continuous with inner face.

Amb (pl. 5, figs. 23–25, 32–34). Amb body outline triangular, dentition consists of transverse lateral plates, longitudinal medial plates; oral apophyse prominent, Adamb notch angular; N very prominent, high, elliptical, pN on prominent ridge, dN slightly inset, forms low ridge. Amb comparisons: Distinguished by triangular amb body, angular Adamb notch, and elongate high prominent N.

Adamb (pl. 5, figs. 19–22). Ossicle outline square, rounded orally; pal angular adradially, psrl moderately high, pa3, pa4 prominent, pm3 large, subquadrate, adoral margin rounded; dm1 deep, spine bases moderate. Adamb comparisons: Angular pal distinguishes *L. magnifica* Adambb from all but those of *L. maculata,* from which they may be distinguished by the presence of a deeper dm1, more prominent spine bases; the pm3 has a rounded oral margin in *L. maculata.*

Pax (pl. 5, fig. 18). Ossicle light, rounded to subquadrate, slightly enlarged tabula, surface of tabula covered by small, densely spaced spinelet bases, column slender, cross section circular, base light with prominent articulation projections.

Material.—One arm fragment, UCMP D-1909; two specimens, USNM 21154, E3099; all from Hawaii, all Recent; ossicle description based on UCMP fragment, R, r unknown, specimen probably large; hypotype no. 10639.

Luidia mascarena Döderlein, 1920
(pl. 6, figs. 1–24)

Luidia mascarena Döderlein, 1920:261–262, pl. 18, fig. 5.

Description.—InfM (pl. 6, figs. 16–24). Ossicle moderately high, outline step-shaped; superambulacral boss prominent; oral side face slightly inclined; abradial region elliptical, long axis horizontal; oral outline linear, oral abradial corner gently rounded; with a transverse row of approximately four moderately large spine bases, bases of various sizes loosely distributed around larger bases, very small bases near face margin; distal articulation ridge moderately long, outline broadly convex, symmetrical, adradial and abradial junctions inclined. InfM comparisons: Distinguished by the symmetrical outline of the distal articulation ridge, the constriction in the outer face, and the prominent superambulacral boss.

Amb (pl. 6, figs. 3–8). Amb body subtriangular, dentition consists of transverse lateral plates, longitudinal medial plates; oral apophyse prominent; Adamb notch subangular; N prominent, elongate, elliptical, pN on low ridge, dN inset. Amb comparisons: Distinguished by subtriangular amb body, subangular Adamb notch, prominent oral apophyse.

Adamb (pl. 6, figs. 9–15). Ossicle outline rectangular, high, ad pm prominent; pal angular adradially, pm1 almost as wide as ossicle, psr1 low, pm3 very small, triangular. Adamb comparisons: Distinguished by moderately angular pal and very small, triangular pm3.

Pax (pl. 6, figs. 1, 2). Spine bearing ossicle proportionally low, massive, column heavy, tabula subrectangular with distinct spine base; base massive, with prominent articulation projections; other Pax with much more slender column, only slightly enlarged circular tabula, prominent articulation projections.

Material.—One specimen, USNM E7255, Bikini Lagoon, Recent; ossicle descriptions based on proximal arm fragment; R = 90 mm, r = 12 mm; hypotype UCMP no. 10641.

Luidia maculata Müller and Troschel, 1842
(pl. 7, figs. 20–30)

Luidia maculata Müller and Troschel, 1842:77–78. Döderlein, 1920:262–266, pl. 18, figs. 4, 13; pl. 19, fig. 16; pl. 20, figs. 23, 24.

Description.—InfM (pl. 7, figs. 23–27). Ossicle moderately high, outline oval; superambulacral boss prominent; oral side face slightly inclined; abradial region rounded, triangular; oral outline convex, oral abradial corner broadly rounded; outer face shorter abradially, with a transverse row of approximately four large spine bases, numerous small-to-moderate spine bases loosely distributed on remainder of outer face, moderately densely spaced small bases near proximal face margin; distal articulation ridge short, outline convex, adradial junction notched, abradial junction inclined; with single elongate ovate articulation process. InfM comparisons: Distinguished by ovate outline and the massive asymmetrical distal articulation ridge.

Amb (pl. 7, figs. 28–30). Amb body subrectangular, dentition consists of transverse lateral plates, longitudinal medial plates; oral apophyse prominent; Adamb notch angular; N very prominent, subcircular, pN on prominent ridge, dN inset, forms low ridge. Amb comparisons: Distinguished by subrectangular amb body, angular Adamb notch.

Adamb (pl. 7, figs. 20–22). Ossicle square, ad pm prominent, pal angular adradially, psr1 moderately high, pm3 large, subquadrate; spine bases small. Adamb comparisons: Distinguished by angular pal and large rectangular pm3.

Pax. Ossicle light, high, tabula slightly enlarged, outline circular-to-rectangular, with densely spaced granule bases; column slender, base light with prominent articulation projections.

Material.—One specimen, USNM 1975, North China Sea, Recent; ossicle descriptions based on proximal arm fragment; R = 70 mm to 110 mm, r = 20 mm; hypotype UCMP no. 10644.

Luidia alternata (Say), 1825
(pl. 6, figs. 25–42)

Asterias alternata Say, 1825:144–145.
Luidia alternata (Say) Lütken: Verrill, 1915:201–203.

Description.—InfM (pl. 6, figs. 37–42). Ossicle high, outline step-shaped; superambulacral boss moderately prominent; abradial region elliptical, long axis vertical; oral outline convex, oral abradial corner broadly rounded; outer face with two large, abradial spine bases, approximately two moderate-sized, oral surface spine bases, small variable-sized spine bases loosely distributed on remainder of face; distal articulation ridge moderately long, outline broadly convex, asymmetrical, adradial ossicle junction gradual, abradial junction inclined. InfM comparisons: distinguished by high step-shape, curved oral outline, the positions of the spine bases, and the outline of the distal articulation ridge.

Amb (pl. 6, figs. 28–36). Amb body triangular; dentition consists primarily of transverse plates; oral apophyse moderately prominent; Adamb notch curved; N prominent, high elliptical, pN on ridge, dN inset. Amb comparisons: Very similar to *L. foliolata*, possibly distinguished by subcircular, less prominent N. Distinguished from Ambb of other species by triangular amb body, moderate oral apophyse, and prominent N.

Adamb (pl. 6, figs. 25–27). Ossicle quadrate, ad pm prominent; psrl moderately high, pm3 high, triangular, oral outlines broadly rounded. Adamb comparisons: Distinguished by the moderately high, triangular pm3 and the rounded oral margin.

Pax. Spine-bearing ossicle massive, tabula large, column stout, base massive, with prominent articulation projections; ossicle without spine base light, tabula small, subcircular, column slender, base with prominent articulation projections.

Material.—Two specimens, UCMP D-1945, from the Gulf of Mexico; numerous specimens, USNM, from Texas, Florida, South Carolina, including 19610, 39837, E5321, E8403; all Recent; ossicle description based on arm of one UCMP specimen, R = 96 mm to 135 mm, r = 15 mm; hypotype no. 10642.

Luidia ludwigi Fisher, 1906
(pl. 5, figs. 1–17)

Luidia ludwigi Fisher, 1906:122–124. Fisher, 1911:113–116, pl. 20, figs. 2, 3; pl. 21, fig. 2; pl. 54, fig. 2.

Description.—Infm (pl. 5, figs. 10–15). Ossicle high, outline step-shaped; superambulacral boss prominent; abradial region broadly elliptical, long axis vertical; oral outline linear, oral abradial corner gently rounded; outer face covered by moderately densely spaced, small-to-moderate-sized spine bases, without distinct size groups; distal articulation ridge moderately long, outline broadly convex, symmetrical, adradial junction gradual, abradial junction gently inclined; usually single, somewhat elongate articulation process, may be disjunct adradially. InfM comparisons: Distinguished by small spine bases, subcircular abradial region, and an unconstricted outer face.

Amb (pl. 5, figs. 6–9, 16, 17). Amb body triangular, dentition consists of transverse lateral plates separated by medial gap; oral apophyse prominent; Adamb notch curved; N prominent, low, elliptical, pN on prominent ridge, dN inset; aboral ridge moderately sharp, adradial part moderately well developed. Amb comparisons: Distinguished by triangular amb body, development of the aboral ridge, and prominent N.

Adamb (pl. 5, figs. 1–5). Ossicle wide, outline rectangular, ad pm rounded, prominent; psrl triangular, very high abradially, pm3 triangular, low; oral margin curved. Adamb comparisons: Distinguished from all Adambb except those of *L. foliolata* by the high psrl and low, triangular pm3; distinguished from *L. foliolata* by a rounded oral margin and somewhat smaller psrl.

Pax. Ossicle low, small, moderately massive, tabula quadrate on marginal ossicle, circular on medial ossicle, column short, massive, base moderately massive, with prominent articulation projections.

Material.—One unnumbered arm fragment, CAS, from the west coast of North America(?); numerous USNM specimens from southern California, including 38479; all Recent; ossicle descriptions based on proximal (?) ossicles of CAS specimen, R, r unknown; hypotype UCMP no. 10640.

Luidia phragma Clark, 1910
(pl. 8, figs. 1–15)

Luidia phragma Clark, 1910:329–330, pl. 2, fig. 1.

Description.—InfM (pl. 8, figs. 10–15). Ossicle moderately high, outline cuneate; superambulacral boss moderately prominent, position strongly adradial; abradial region elliptical, long axis horizontal; oral outline gently rounded, oral abradial corner broadly rounded; outer face with slight abradial constriction, with three large abradial spine bases, three moderate-sized oral surface bases, moderately densely spaced small-to-moderate spine bases on remainder of surface; distal articulation ridge short, outline triangular, adradial ossicle junction very gentle, abradial junction perpendicular; proximal articulation ridge corner slightly obtuse; articulation process elongate, ovate, may be partially disjunct adradially. InfM comparisons: Distinguished by gentle inner face step, adradially placed superambulacral boss, and triangular outline of the distal articulation ridge.

Amb (pl. 8, figs 4–9). Amb body subtriangular, dentition consists of transverse lateral plates, with small medial gap; oral apophyse prominent; N moderately prominent, elongate, elliptical, pN prominent, dN slightly inset; aboral ridge moderately sharp, well developed adradially, aboral ridge distinct from N. Amb comparisons: Most similar to Ambb of *L. tesselata*; *L. phragma* Ambb with more triangular amb body, aboral ridge more prominent adradially.

Adamb (pl. 8, figs. 1–3). Ossicle square, ad pm moderately prominent; psr1 height moderate, pm3 moderately high, triangular; with small oral flange; dm1 deep. Adamb comparisons: Distinguished by the moderately high psr1, moderate pm3, and square outline.

Fst Pax. Ossicle moderately light, tabula rectangular, with small, densely spaced granule bases; column triangular, grades into tabula; base with prominent articulation projections, especially abradially; Pax similar but lighter, tabula quadrate on ossicle near arm margin, circular on medial ossicle.

Material.—Three specimens, UCMP D-1943; numerous specimens, USNM, including 40015, 40016, all from Baja California; all Recent; ossicle descriptions based primarily on one UCMP specimen, R = 86 mm, r = 15 mm; hypotype no. 10645.

Luidia foliolata Grube, 1866

(pl. 9, figs. 1–26)

Luidia foliolata Grube, 1866:59. Fisher, 1911:106–113, pl. 19, figs. 1–3; pl. 21, figs. 3–5; pl. 54, fig. 3.

Description.—InfM (pl. 9, figs. 12–20). Ossicle moderately high, outline step-shaped; superambulacral boss very prominent; abradial region subcircular; oral outline linear, abradial corner gently rounded; outer face constricted abradially, with two large, abradial spine bases, remainder of surface with small-to-moderate-sized, moderately densely spaced spine bases, small bases dense along ossicle margins; distal articulation ridge long, outline sharply convex, asymmetrical, adradial junction gradual, abradial junction perpendicular, with one large, subcircular articulation process. InfM comparisons: Distinguished by very long, asymmetrical, distal articulation ridge, very prominent superambulacral boss, and subcircular abradial region.

Amb (pl. 9, figs. 9–11, 21–26). Amb body triangular, dentition consists of transverse lateral plates, vertical medial plates; oral apophyse moderately prominent; Adamb notch curved; N very prominent, high, elliptical, pN on ridge, dN inset. Amb comparisons: Distinguished by triangular amb body and moderately prominent oral apophyse; very similar to Ambb of *L. alternata*, N of *L. foliolata* are lower, more elliptical.

Adamb (pl. 9, figs. 2–8). Ossicle rectangular, wide, ad pm prominent, rounded; psr1 triangular, very high; pa3, pa4 very prominent; pm3 low, triangular. Adamb comparisons: Distinguished by the very high triangular psr1 and the small, triangular pm3.

Pax (pl. 9, fig. 1). Ossicle relatively light, low; tabula quadrate on marginal ossicle, subcircular or irregular on medial ossicle, with densely spaced granule bases; tapers into short or moderately long, subcircular, light, massive column; base moderately massive, with prominent articulation projections; Fst Pax similar, larger.

Material.—About 15 Recent specimens, UCMP D-1944, all from the Pacific Coast of North America (?); one late Pliocene or early Pleistocene specimen, UCMP D-195, from south of San Francisco; numerous Recent USNM specimens from the Pacific coast of North America, including 31455–31458; ossicle descriptions based primarily on UCMP specimens; hypotype no. 10647.

Remarks.—The fossil specimen that has been assigned to this species is imbedded in a highly indurated beach cobble. The specimen was preserved virtually intact; inferomarginal spines are still in place, and spinelets are present on the paxillae. The specimen is assigned to *L. foliolata* because of the InfM outline. The abradial region is high; the inner face step is distinct; the spine bases correspond in position to those of Recent specimens of *L. foliolata*. In addition, the outline of the

Adambb and the development of the Pax is, as far as can be determined, very similar to that of Recent specimens of *L. foliolata.*

Luidia etchegoinensis, n. sp.
(pl. 12, figs. 1–20)

Diagnosis.—Alternata Group *Luidia*, InfM outline ovate, oral surface outline linear, spine bases prominent; inner face step gentle, superambulacral boss weakly developed, distal articulation ridge long, symmetrical; amb body massive, long, strongly triangular, N very low; Adamb pentagonal, with prominent medial spine bases, abruptly tapering pm1, psr1; Pax massive, bear massive, densely spaced, tablet-shaped granules.

Description.—InfM (pl. 12, figs. 13–20). Ossicle moderately high, outline ovate; inner face step gentle, superambulacral boss not prominent; oral side face continuous with inner face; abradial region subelliptical, long axis horizontal; oral outline linear, oral abradial corner gently rounded; outer face with about four large spine bases, on oral surface; distal articulation ridge length moderate, outline symmetrical, broadly convex, adradial ossicle junction gradational, abradial junction inclined. InfM comparisons: Distinguished by ovate outline, spine base size and development, elliptical abradial region, gentle inner face step, and the length and shape of the distal articulation ridge.

Amb (pl. 12, figs. 9–12). Amb body outline triangular, dentition consists of transverse lateral plates, medial gap present; oral apophyse moderately prominent; Adamb notch curved; N prominent, very low, pN with a distinct elongate groove. Amb comparisons: Distinguished by long, strongly triangular massive amb body, elongate low N, and low, rounded aboral ridge.

Adamb (pl. 12, figs. 2–8, 12). Ossicle large, massive, outline pentagonal, ad pm moderately prominent; pm1 high, about ¾ ossicle width, tapering abruptly abradially, psr1 high abradially; pa3, pa4 very prominent, pm3 triangular, high; da2 well developed, dm1 deep; outer face with prominent medial spine bases, no prominent bases adradially, abradially. Adamb comparisons: Distinguished by ossicle shape, spine base development, and the shape of the pm1 and psr1.

Material.—One arm fragment, about 33 mm in length, UCMP B-5046, Pliocene, Etchegoin Formation, Kettleman Hills, east of Avenal, California; holotype no. 10652.

Remarks.—The single row of marginals and the morphology of all ossicles are typical of the genus *Luidia.* The InfMM, with their broad distal articulation ridges, irregular outer faces, and prominent spine bases, the Ambb, with the massive asymmetrical triangular bodies and low aboral ridges, and the Adambb, with their prominent pm1 and high psr1, are all typically Alternata Group-like in development; the paxiallae, though somewhat large, do not appear to be significantly larger than those of *L. ludwigi*—certainly not significantly larger than those of *L. foliolata.* The large paxillae would serve to place this specimen in *Armaster*, that subgenus described by Döderlein as having little or no reduction of the side paxillae. Adambb are generally conservative in evolution and often not obviously distinct, even between groups; those of *L. etchegoiensis* are distinct from those of other species. Both the InfMM and Ambb also appear distinct in those features that are here considered to vary between species.

Luidia sanjoaquinensis, n. sp.
(pl. 12, figs. 21–30)

Diagnosis.—Alternata Group *Luidia* with step-shaped InfM, inner face step gentle, oral outline broadly rounded, superambulacral boss weakly developed, spine bases prominent, bearing broad, paddle-shaped spines; Amb with high N, prominent oral apophyse, subangular Adamb notch; Adamb pentagonal, with prominent medial spine bases and a large, triangular, rapidly tapering pm1; Pax massive, bearing globular granules.

Description.—InfM (pl. 12, figs. 25–28). Ossicle moderately high, outline step-shaped; inner face step moderate, superambulacral boss not prominent; oral side face continuous with inner face; abradial region subcircular; oral outline broadly curved, oral abradial corner broadly rounded; with a transverse row of approximately five large spine bases; distal articulation ridge length moderate, outline broadly convex, adradial ossicle junction gradational, abradial junction inclined. InfM comparisons: Distinguished by outline, distribution of spine bases, gentle inner face step, and weakly developed superambulacral boss.

Amb (pl. 12, figs. 21, 23). Amb body outline probably triangular; oral apophyse prominent; Adamb notch subangular, N prominent, moderately high; aboral ridge low, rounded. Amb comparisons: Amb preservation is poor, but the shape of the N combined with the outline of the Adamb notch appears distinctive.

Adamb (pl. 12, figs. 22, 24). Ossicle outline pentagonal, ad pm prominent; pm1 large, triangular, rapidly tapering; psr1 very high, pa3, pa4 prominent; pm3 low, triangular; medial spine bases prominent. Adamb comparisons: Distinguished by the pentagonal ossicle outline, the shape of the pm1 and psr1, and the spine base distribution.

Pax (pl. 12, figs. 29, 30). Ossicle moderately large, massive; tabula quadrate, column circular or elliptical, base with prominent articulation projections; tabula covered by globular granules, a central granule surrounded by a ring of about six similar granules; spines not present.

InfM Spines (pl. 12, fig. 28). InfMM bear at least two abradial short, broad, paddle-like spines with very blunt tips.

Material.—One arm fragment, about 40 mm in length, UCMP D-2439, early Pleistocene, San Joaquin Formation, Kettleman Hills, east of Avenal, California; holotype no. 10653.

Remarks.—The single row of marginals and the morphology of all ossicles are typical of the genus *Luidia*. The specimen is assigned to the Alternata Group because of the outline of the InfMM, their broad articulation ridges and irregular outer faces, the prominent spine bases, and the low, rounded aboral ridge of the Amb. The combination of features on the three principal ossicle types and the granules of the paxillae and the paddle-shaped spines of the InfMM serve to distinguish *L. sanjoaquinensis* from other species.

Luidia sp. A

(pl. 13, figs. 27, 28)

Description.—InfM. Ossicle wide, moderately massive, may have had moderately prominent spine bases, abradial region of ossicle high, adradial region very wide, low, inner face step sharp, inner face surface convex.

Pax. Two (possibly one) rows of small paxillae with circular-to-subquadrate tabulae immediately aboral to InfMM; next, one row of Pax of variable size, some very large with quadrate tabulae; next, about three rows of generally regularly arranged ossicles, decreasing in size, angularity of tabulae outline nearer arm axis; central area with many small Pax with rounded tabulae; paxillae spinelets possibly small, granular.

Material.—One fossil arm fragment, largely recrystallized, about 48 mm long; Miocene, USGS M-3917; hypotype USNM no. 651560.

Remarks.—The paxillae pattern described above is one that occurs in many Alternata Group members; the InfM appears most like those of members of this group; it is therefore tentatively considered to be an Alternata Group member. InfM morphology and paxillae size and distribution suggest that the specimen represents an unknown species. Sufficient information is not available, however, to warrant erection of a new species.

Luidia sp. B

(pl. 13, figs. 23, 24)

Description.—InfM. Ossicle light, high, spine bases moderately prominent, articulation ridges wide; Amb with very wide triangular amb body, very prominent pN, short deep Unt G, well developed oral groove.

Adamb: pm3 deep, ad pm weak; spine bases large, prominent; distal ossicles proportionately narrower.

Pax. Small, some with well developed spine-bearing (?) circular tabulae, slender tapering columns, others with small cylindrical columns, tabulae weak, bases light with prominent articulation projections.

Material.—External molds of arm fragments of a fossil specimen, length of largest fragment 90 mm; primarily oral surfaces of oral ossicles preserved, aboral portions of InfMM, Ambb, Adambb lacking; from the Pliocene Pico Formation; from about 35 miles east of Santa Barbara, California; UCMP A-415; hypotype no. 10662.

Remarks.—These specimens can be placed in the Alternata Group of the genus *Luidia* on the basis of large, massive Amb ossicles with large, triangular amb bodies; large, rectangular Adambb with prominent deep dmls; high, light InfMM with long articulation ridges; and small, light Pax. General ossicle morphology appears distinct from that of known species, but preservation is not complete enough to warrant description of a new species.

Luidia sp. C
(pl. 13, figs. 25, 26)

Description.—InfM (pl. 13, fig. 26). Ossicle relatively wide; outer face short; with about two large spine bases near oral abradial corner, about five moderate-sized bases closely spaced along oral margin, larger bases offset slightly distally from center line of face, remainder of surface with smaller bases; oral abradial corner rounded; abradial region of ossicle high, adradial region very wide, low; inner face step sharp, surface convex.

Material.—One fossil arm fragment, recrystallized, from the Pliocene "Etchegoin" Formation; from about 20 miles southeast of San Luis Obispo, California; fragment about 33 mm long; UCMP A-1461 hypotype no. 10654.

Remarks.—The distinct inner face step, the high abradial region, and the moderate-size spine bases suggest the Alternata Group. InfM morphology suggests a new species, but, because of lack of sufficient information, it is not named. InfM morphology suggests that this specimen and *Luidia* sp. B may represent closely related species.

Species of the Quinaria Group of Döderlein, 1920

The following features occur in Quinaria Group species, except as modified under individual descriptions.

Inferomarginal ossicles: ossicle high, massive, inner face step gentle, oral side face inclined to inner face; distal articulation ridge outline gently convex, adradial junction notched, abradial junction inclined; proximal articulation ridge angle slightly obtuse, vertical component linear; with single L-shaped articulation process.

Ambulacral ossicles: Ossicles massive, amb body rectangular; dentition variable, consists of longitudinal and transverse plates; Adamb notch curved; aboral ridge distinctly set off, broadly rounded, adradial part prominent.

Luidia quinaria v. Martens, 1865
(pl. 7, figs. 1–19)

Luidia maculata Müller and Troschel var. *quinaria* v. Martens, 1865:352–353.
Luidia quinaria: Döderlein, 1920:275–277, pl. 20, fig. 26.

Description.—InfM (pl. 7, figs. 14–19). Ossicle outline ovate; superambulacral boss moderately prominent; abradial region outline subcircular; oral abradial corner broadly rounded; outer face long, parallel-sided, with abradial constriction; with small-to-moderate-sized, moderately densely spaced spine bases, not sharply differentiated in size; with numerous small spine bases along ossicle margin; distal articulation ridge short. InfM comparisons: Distinguished by its outline and elongate articulation process.

Amb (pl. 7, figs. 8–13). Oral apophyse moderate, GG subequal, margins parallel, pGG tapers abradially, dGG tapers adradially; junction between pGG and ossicle smooth; N prominent, elliptical, high, pN on ridge, dN inset. Amb comparisons: Distinguished by massive form, subequal GG.

Adamb (pl. 7, figs. 3–7). Ossicle massive, outline irregular, angular, ad pm prominent; pa1, pm1 large, psr1 very high, triangular, pa4 inflated; pm3 irregular, triangular, with oral flange; dm1 small. Adamb comparisons: Similar only to Adambb of *L. penangensis; L. quinaria* Adambb have a smaller oral surface.

Pax (pl. 7, figs. 1, 2). Ossicle short, massive; tabula quadrate on abradial ossicle, circular on medial ossicle, with densely spaced granule bases; column stout; base fairly massive, with prominent articulation projections.

Material.—One unnumbered specimen, CAS, from Japan; several USNM specimens, all from Japan, including 38689; all Recent; ossicle descriptions based on proximal and medial ossicles of one arm of CAS specimen, R = 100 mm, r = 19 mm, hypotype UCMP no. 10643.

Luidia penangensis de Loriol, 1891

(pl. 8, figs. 16–33)

Luidia penangensis de Loriol, 1891:24, pl. 3, figs. 2, 3. Döderlein, 1920:282–283, pl. 18, fig. 12; pl. 20, fig. 32.

Description.—InfM (pl. 8, figs. 22–27). Ossicle outline cuneate; superambulacral boss prominent; abradial region high, angular, triangular; oral outline linear, oral abradial corner angular; outer face long, parallel-sided; with three large spine bases on oral surface, small spine bases along ossicle margins, few moderate-sized spine bases between large bases; distal articulation ridge length moderate. InfM comparisons: Distinguished by angular oral abradial margin, prominent spine bases, and superambulacral boss.

Amb (pl. 8, figs. 28–33). Oral apophyse prominent; GG asymmetrical, junction between pGG and ossicle smooth; N prominent, elliptical, low, pN on prominent ridge, dN on small ridge. Amb comparisons: Similar only to *L. quinaria* Ambb, *L. penangensis* Ambb are distinguished by their asymmetrical GG.

Adamb (pl. 8, figs. 17–21). Ossicle massive, outline irregular, angular, ad pm prominent; pa1 large, psr1 large, triangular; pa3, pa4 prominent; pm3 irregular, triangular; dm1 small; outer face large. Adamb comparisons: Similar only to Adambb of *L. quinaria;* the outer face of *L. penangensis* Adambb are proportionately larger.

Pax (pl. 8, fig. 16). Ossicle massive, tabula outline irregular, many with spine bases, very short massive column, very large massive base with prominent articulation projections; Fst Pax similar, larger.

Material.—One specimen, USNM E6637, from the South Malacca Straits, Recent; R = 95 mm to 140 mm, r = 20 mm; ossicle descriptions based on proximal ossicles of one arm; hypotype UCMP no. 10646.

Species of the Ciliaris Group of Döderlein, 1920

The following features occur in Ciliaris Group species, except as modified under individual descriptions.

Inferomarginal ossicles: Outline crescentic, without a distinct inner face step; superambulacral boss moderately prominent; distal articulation ridge parallel to ossicle curvature, outline rectangular, very wide, very long, with two marginal articulation processes; proximal ridge short, wide.

Ambulacral ossicles: Amb body triangular, asymmetrical; dentition weak, with medial gap; Unt G small, oral apophyse prominent; GG subsymmetrical, N low, elongate, triangular, aboral ridge low, rounded.

Luidia ciliaris (Philippi), 1837
(pl. 10, figs. 1–26)

Asterias ciliaris Philippi, 1837:194.
Luidia ciliaris: Döderlein, 1920:287–288, pl. 18, fig. 8; pl. 19, fig. 17; pl. 20, fig. 34.

Description.—InfM (pl. 10, figs. 19–26). Ossicle light; superambulacral boss moderately prominent, paxilla groove well developed; oral surface outline curved, outline of outer face cuneate, with three moderate-sized spine bases; distal articulation ridge outline very angular, with distinct marginal notch. InfM comparisons: Distinguished by light form, cuneate outer face outline, and angular distal articulation ridge.

Amb (pl. 10, figs. 10–18). Amb body triangular, outline similar to Alternata Group species, dentition moderate, medial gap moderate, Adamb notch angular, pN prominent, dN inset, aboral ridge not distinctly defined adradially. Amb comparisons: Distinguished by Alternata Group-like amb body, prominent dentition.

Adamb (pl. 10, figs. 3–9). pa1 angular adradially, pm1 moderately wide, psr1 relatively large, spine bases weak, dm1 large. Adamb comparisons: Distinguished by angular pa1, relatively large dm1.

Pax (pl. 10, figs. 1, 2). Ossicle high, tabula semicircular, column slender, base with prominent articulation projections; Fst Pax low with semicircular tabula, short stout column, light base with prominent articulation projections.

Material.—One unnumbered specimen, HMS, from Naples, Italy; one specimen, USNM 3840, Kieler Bucht, Germany; both Recent; ossicle descriptions based on ossicle of one arm of HMS specimen; hypotype UCMP no. 10648.

Luidia asthenosoma Fisher, 1906
(pl. 11, figs. 1–23)

Luidia asthenosoma Fisher, 1906:124–127. Fisher, 1911:116–119, pl. 20, fig. 1; pl. 21, fig. 1; pl. 54, fig. 1.

Description.—InfM (pl. 11, figs. 16–23). Ossicle moderately light, paxilla groove well developed; oral surface outline curved, outer face parallel-sided; two moderately large spine bases on oral surface, remainder of surface with smaller loosely spaced spine bases, most dense near ossicle margins; distal articulation ridge with distinct marginal notch. InfM comparisons: Distinguished by parallel outer face margins, numerous spine bases.

Amb (pl. 11, figs. 10–15). Amb body triangular, medial gap large; Unt G small, Adamb notch subangular; pN on very prominent ridge, dN on low ridge, adradial part of aboral ridge prominent. Amb comparisons: Distinguished by subangular adamb notch, small Unt G.

Adamb (pl. 11, figs. 3–9). Ossicle very high, ad pm prominent; pm1 narrow, psr1 moderate, pm3 small, triangular, with large oral flange; spine bases moderate. Adamb comparisons: Distinguished by prominent ad pm, narrow pm1, oral flange on pm3.

Pax (pl. 11, figs. 1, 2). Ossicle light, tabula small, slender, column slender, articulation projections prominent; Fst Pax low, fairly massive, tabula elongate, notched for InfM, column short, massive, base light, with prominent articulation projections.

Material.—One specimen, UCMP D-1944, Pacific coast of North America (?); several USNM specimens, including 21929; ossicle descriptions based on arm of UCMP specimen; hypotype no. 10650.

Luidia elegans Perrier, 1876
(pl. 10, figs. 27–52)

Luidia elegans Perrier, 1876:256–257. Verrill, 1915:203–205, pl. 16, figs. 4, 4a; pl. 19, fig. 1.

Description.—InfM (pl. 10, figs. 44–52). Ossicle moderately heavy, outline crescentic, truncate adradially; superambulacral boss moderately prominent, truncate adradially; paxilla groove weakly developed, oral surface outline angular, outer face outline parallel-sided; with three moderate-sized spine bases, spine base near aboral margin smallest; small spine bases irregularly distributed, most dense near ossicle margins. InfM comparisons: Distinguished by the sharply truncate adradial margin.

Amb (pl. 10, figs. 35–43). Amb body triangular, subsymmetrical, N on prominent ridges, similar; medial gap moderate, Adamb notch curved, adradial part of aboral ridge prominent. Amb comparisons: Distinguished by subsymmetrical amb body.

Adamb (pl. 10, figs. 29–34). pm1 moderately wide, psr1 triangular, extends to aboral margin; pm3 triangular, with oral flange; spine bases large. Adamb comparisons: Distinguished by large psr1, prominent spine bases.

Pax (pl. 10, figs. 27, 28). Ossicle slender, light, base with prominent articulation projections; Fst Pax low, massive, tabula slightly enlarged, circular to rectangular, column short, massive, base massive.

Material.—Fragments of several USNM specimens, from east coast of North America; Recent; ossicle descriptions from proximal ossicles of one specimen, USNM 9376; hypotype UCMP no. 10649.

Luidia neozelanica Mortensen, 1925

(pl. 11, figs. 24–44)

Luidia neozelanica Mortensen, 1925:278–281, pl. 12, fig. 5; text figs. 5, 6a–6d.

Description.—InfM (pl. 11, figs. 33–39). Ossicle massive, paxilla groove weakly developed, oral surface outline curved, outline of outer face cuneate; outer face with two or three large spine bases, those with three have one near aboral margin; moderate-to-small-sized spine bases irregularly distributed on remainder of face, many small bases along face margin; articulation processes elongate, nearly fused, articulation ridge margin lacks notch. InfM comparisons: Distinguished by prominent spine bases, elongate articulation process without marginal notch; cuneate outer face outline.

Amb (pl. 11, figs. 40–44). Amb body triangular, medial gap large; oral apophyse moderately prominent, Adamb notch curved; GG asymetrical, pN on very prominent ridge, triangular, dN on low ridge, adradial part of aboral ridge prominent. Amb comparisons: Distinguished by asymmetrical GG, moderately prominent oral apophyse, curved Adamb notch.

Adamb (pl. 11, figs. 28–32). pm1 moderately wide, psr1 moderately large, pm3 subquadrate, with oral flange; spine bases large. Adamb comparisons: Distinguished by large spine bases, subquadrate pm3.

Pax (pl. 11, figs. 24–27). Ossicle small, tabula circular, not sharply set off from slender column, base slender with prominent articulation projections; Fst Pax tabula elliptical, with deep abradial indentation for InfM, column very weakly developed, large flat bases formed largely of articulation projections.

Material.—Arm fragments of one specimen, USNM E10,044, from New Zealand; Recent; R at least 110 mm; ossicle descriptions from proximal ossicles of one arm; hypotype UCMP no. 10651.

Family ASTROPECTINIDAE

Genus *Astropecten* Gray, 1840

The west American astropectens are varied in morphology and apparently closely related. This has led to differing taxonomic interpretations among the various authors who have studied these starfish. Fisher (1911) recognized three species: *A. armatus, A. californicus,* and *A. ornatissimus.* He says of *A. armatus,* "This species is so variable that it is difficult to make its positive characters intelligible through the median of description" (ibid., p. 60). He further noted (ibid., p. 67) that, "Until an extensive series of Mexican and southern Astropectens is compared by one man it will not be possible to determine the validity of the rather numerous nominal species." This still has not been done. In the present study, the species concepts of Fisher have been used, because they were most readily applied in the collections available.

Döderlein (1917), who divided the genus into sixteen "groups" in a manner similar to his division of *Luidia,* places representatives of these three species in his

Brasiliensis Group. The complexity and continuity of morphology observed in complete specimens of west American species of *Astropecten* of the Brasiliensis Group also occurs in their ossicles. For this reason, and also because specimens of only three species belonging to this group were available, the three species are first considered together under the Brasiliensis Group and then are diagnosed separately.

Species of the Brasiliensis Group of Döderlein, 1917

SupM.—Ossicle high, narrow, approximately triangular in side face outline; outer face margins parallel, rounded, outer surface flat or gently arched, covered by densely spaced granule bases, which diminish in size near surface margins; side faces similar, fasciolar surfaces large, covered laterally by fine fasciolar spinelet bases; articulation area triangular; articulation ridges well developed except along oral margin; aboral side face small; inner face smooth; intermarginal face concave, distal ossicles asymmetrical, proportionately low; interbrachial ossicles cuneate.

InfM.—Ossicle massive, low, wide, asymmetrical, triangular-to-oval in side face outline; outer face margins parallel, angular; oblique series of approximately seven dissected spine bases of variable size along distal margin, largest two bases near abradial margin, second row of small bases may be present proximal to large row; remainder of surface covered by moderately densely spaced spine bases, smaller near face margins; side faces similar, fasciolar areas large, covered laterally by fine fasciolar spinelet pits; articulation ridges well developed, especially orally, may be disjunct; intermarginal face, inner face and oral side face continuous; articulation processes, superambulacral bosses, interactinal ossicle contacts not prominent; interbrachial ossicles cuneate, wide.

Amb.—Ossicles high, short; amb body outline rectangular; dentition well developed, consists of vertical medial, horizontal lateral plates; Unt G, oral groove well developed, deep; oral apophyse very weak; pGG weak, U-shaped, opens orally; N high, prominent, ovate; aboral ridge high, relatively sharp; superambulacral contact prominent; proximal ossicles proportionately high, short.

Adamb.—Side face outline rectangular; pa1 cuneate, pm1 high, narrow, pm2 large; pa3, pa4 prominent, pm3 large, triangular; da2 prominent, dm1 large, quadrate; many moderately prominent spine bases; near-oral ossicles short, high, features very distorted.

Pax.—Ossicle simple, tabula small, rounded, articulation projections simple.

Astropecten armatus Gray, 1840
(pl. 14, figs. 36–55)

Astropecten armatus Gray, 1840:181. Fisher, 1911:56–61, pl. 5, figs. 1, 2; pl. 7, figs. 3, 6; pl. 50, fig. 4; pl. 51, fig. 3.

Description.—Superomarginals with spine bases on at least proximal ossicles, bases located near aboral margin; inferomarginals with a row of about four large, dissected spine bases curving from proximal aboral corner toward the abradial end of the distal margin of the outer face, second and third bases from aboral largest; two or three additional enlarged bases spaced along distal margin of oral face; one or two very slightly enlarged bases may be present abradially proximal to the enlarged bases; proximal InfM outline triangular; dGG of Amb subrectangular in outline.

Material.—Several specimens, UCMP D-1956, Balboa Island, California; numerous CAS specimens, numerous USNM specimens including 8961, 39806, California; all Recent; ossicle descriptions based on UCMP specimens; hypotype no. 10669.

Astropecten californicus (Fisher), 1906
(pl. 14, figs. 1–12)

Astropecten californicus Fisher, 1906:299–300. Fisher, 1911:61–67, pl. 6, figs. 1, 2; pl. 7. fig. 1; pl. 50, fig. 5; pl. 51, fig. 2.

Description.—Inferomarginals with a row of about two to four dissected spine bases curving from proximal aboral corner toward abradial end of the distal margin of the outer face; two distinctly enlarged bases always present, others, if present, are only slightly enlarged; about two additional enlarged spine bases along distal margin of oral face, a second row of two or three very slightly enlarged bases may occur proximal to the first row; bases relatively smaller than those of *A. armatus, A ornatissimus;* InfM outline triangular; dGG of Amb subcircular in outline.

Material.—Numerous specimens, CAS, and USNM, including 30786, California; Recent; ossicle descriptions based on arm of one CAS specimen of R = 47 mm, r = 13 mm; hypotype UCMP no. 10666.

Astropecten ornatissimus Fisher, 1906

(pl. 14, figs. 25–35)

Astropecten ornatissimus Fisher, 1906:119–121. Fisher, 1911:67–71, pl. 6, figs. 3, 4; pl. 7, fig. 2; pl. 51, fig. 1.

Description.—Superomarginals relatively low, with a distinct aboral side face; inferomarginals with three to five large, dissected spine bases curving from proximal aboral corner toward the abradial end of the distal margin of the outer face, second and third bases from aboral largest; one to three additional bases spaced along distal margin; large row of bases usually paralleled proximally by a second row of about two smaller bases, InFM outline ovate, Amb with very high, frequently serrate aboral ridge.

Material.—Three specimens, CAS, numerous specimens, USNM 38805, California; Recent; ossicle descriptions based on arm of specimen of R = 68 mm, r = 15 mm; hypotype UCMP no. 10668.

Species of the Marginatus Group of Döderlein (1917)
Astropecten regalis Gray, 1840
(pl. 14, figs. 13–24)

Astropecten regalis Gray, 1840:181. Döderlein, 1917:110; pl. 3, fig. 6; pl. 11, figs. 8, 9.

Description.—SupM (pl. 14, figs. 16, 17). Ossicles similar to those of the Brasiliensis Group; proximal ossicle triangular, rounded in side face outline, medial ossicle relatively wider, subrectangular in outline; outer face margins parallel, angular; surface with about five fairly large weakly developed spine bases along distal margin, remainder of surface covered by small, faint, densely spaced granule bases; side faces similar, fasciolar surface moderately large becoming proportionately smaller on distal ossicles, lateral part of surface covered by small fasciolar spine bases; articulation area prominent, triangular, articulation ridges well developed except orally; aboral side face well developed, intermarginal face with weakly developed U-shaped articulation process. SupM comparisons: Distinguished by spine base pattern, development of aboral side face, and relative proportions.

InfM (pl. 14, figs. 22–24). Ossicle massive, quadrate, low; outer face margins parallel, angular, surface flat; two large U-shaped spine bases at abradial extremity, closed margin of U inflected slightly proximal of oral, longitudinal row of fine spinelet bases aboral of large spine base pair, irregular slightly oblique row of two to four spinelet bases oral of large spine base pair; transverse row of about four moderate-sized spine bases along distal margin of oral surface, two or three slightly enlarged spine bases may occur along proximal margin of face, remainder of surface covered by small, moderately densely spaced granule bases; fasciolar surface rather small, very low adradially, with fasciolar spinelet bases abradially; articulation ridges well developed except aborally, may be disjunct; superambulacral bosses moderately prominent, parts of more than one occurring on an ossicle; intermarginal face with weakly developed U-shaped articulation process opening adradially. InfM comparisons: Distinguished by general proportions and spine base development.

Amb (pl. 14, figs. 18–21). Ossicle proportionately wide; amb body outline rectangular, dentition well developed, consists of vertical medial, horizontal lateral plates; Unt G, oral groove deep; oral apophyse weak; pGG weak, U-shaped, opens orally, dGG large, subrectangular; aboral ridge high, sharp, superambulacral boss well developed. Amb comparisons: Distinguished by shape of large GG, superambulacral boss.

Adamb (pl. 14, figs. 13–15). Side face outline rectangular; pal large, cuneate, pm1 high, narrow, pa2 moderately prominent, pm2, pm3, pa3 prominent, well developed; dm1, da2 prominent. Adamb comparisons: Distinguished by relatively wide proportions.

Pax. Ossicle small, tabula small, column proportionately long, base swollen, elongate in direction of arm axis with slightly forked termini, shorter unforked lateral projections.

Material.—Two specimens, UCMP A-8122, two USNM specimens, 39889, 39895, all from Mexico; all Recent; ossicle descriptions based on some proximal and medial ossicles of one arm of one UCMP specimen R = 50 mm, r = 16 mm; hypotype, no. 10667.

Astropecten sp.
(pl. 13, figs. 1–22)

Astropecten (?) Zullo et al., 1964:334.

Description.—Typically astropectenoid, with slender, gently tapering arms, acute interbrachial angles, and a relatively small disc; R of largest specimen approximately 45 mm. Sup M (pl. 13, figs. 13–16, 22). Proximal ossicle high, narrow, outline rectangular; outline of distal ossicle square, asymmetrical outer face essentially continuous with fasciolar surfaces of side face; without spine bases; articulation areas triangular, relatively small on distal ossicles, articulation ridges well developed except orally; aboral side face not distinct; inner face smooth; intermarginal face convex; interbrachial ossicles cuneate. SupM comparisons: Distinguished by lack of spine bases, very rounded margins, square outline, and small articulation area on distal ossicles.

InfM (pl. 13, figs. 17–22). Ossicle massive, low, wide, asymmetrical, side face outline oval; outer face flat or slightly convex, margins angular; with a row of about five enlarged closely spaced spine bases curving from the middle of the aboral edge of the ossicle to the abradial distal margin of the oral surface, middle bases of row slightly larger than end bases; all other spine bases subequal in size; two or three enlarged bases occur spaced along the distal margin of the ossicle, a second row of several very slightly enlarged bases is present proximal of the marginal row; remainder of the surface covered by moderately densely spaced spine bases; side faces similar; articulation areas large, articulation ridges well developed; inner face with small superambulacral bosses, intermarginal face concave. InfM comparisons: Distinguished by spine base distribution.

Amb (pl. 13, figs. 10–12, 21). Amb body long, subtriangular, massive, symmetrical; Unt G, oral groove well developed, deep, oral apophyse weak; pGG weak, U-shaped, opens orally; dGG large; N prominent, ovate; aboral ridge relatively low, massive, rounded; superambulacral contact prominent. Amb comparisons: Distinguished by massive form, triangular amb body, and rounded aboral ridge.

Adamb (pl. 13, figs. 3–9). Ossicle outline rectangular, ad pm prominent; pm1 small, narrow, pm2 large, pa4, da2 prominent; outer surface large, flat; surface with a furrow margin ridge, two rows of three discrete spine bases, and a single abradial base. Adamb comparisons: Distinguished by relatively large, flat oral surface and spine base development.

Pax (pl. 13, figs. 1, 2). Ossicle small, columnar, tapers medially, expands aborally, orally.

Material.—Fragments of about twelve severely leached specimens from the Lower Oligocene Keasey Formation near the town of Mist, in northwest Oregon; R. about 45 mm, r about 12 mm; UCMP A-5018; hypotype no. 10656.

Discussion.—Generic assignment: The general form is astropectenid; the disc small, interbrachial angles sharp, and arms rapidly tapering with massive marginals. Both series of marginals are very similar to those of members of the Brasiliensis Group. Ambb and Adambb are generally Astropectenid. *A.* sp. is not assigned to a group or a species because ossicle morphology of Recent astropectens has not been adequately studied.

Genus *Psilaster* Sladen, 1885
Psilaster pectinatus (Fisher), 1905
(pl. 15, figs. 12–24)

Bathybiaster pectinatus Fisher, 1905:295–296.
*Psilaster pectinatus:*Fisher, 1911:72–78; pl. 7, figs. 5, 7; pl. 10, figs. 1–3; pl. 50, fig. 3.

Description.—SupM (pl. 15, figs. 22–24). Ossicle quadrate in side face outline, triangular in aboral outline; outer face short, continuous with fasciolar surfaces of the side faces and aboral surface of ossicle, aboral surface arched; outer face with one small spine base near aboral margin, moderate-sized dissected spine base oral to small base; remainder of surfaces covered by fine fasciolar spinelet bases; articulation surface small, triangular, articulation ridges well developed except orally; intermarginal face bordered by moderately prominent articulation ridges. SupM comparisons: Proximal ossicles distinctive in overall proportions; distal ossicles similar to those of *Astropecten* but retain relatively small articulation areas.

InfM (pl. 15, figs. 18–21). Ossicle asymmetrical, triangular in side face and oral outline; outer face and fasciolar portion of side faces continuous; outer face with two or proximally three approximately vertically aligned large dissected spine bases, remainder of surface covered by moderately densely spaced moderate-sized spine bases; articulation areas small, elliptical, articulation ridges well developed except aborally, intermarginal face with well developed articulation ridge. InfM comparisons: Distinguished by general form, small articulation areas.

Amb (pl. 15, figs. 15–17). Ossicle moderately high, amb body long, triangular, dentition well developed, consists of horizontal marginal plates, vertical medial plates; Unt G and oral groove well developed, proximal margin of Unt G extends orally to form ridge; oral apophyse weakly developed; GG triangular, of similar size, pGG open U shape, tapers abradially, dGG tapers adradially; N ovate, high, pN forms distinct ridge; aboral ridge low, smooth, superambulacral contact prominent; distal ossicles similar, except amb body is relatively more prominent. Amb comparisons: Distinguished by the long, massive amb body and triangular GG.

Adamb (pl. 15, figs. 12–14). Ossicle high, outline irregular; pm1 about ⅔ ossicle width, triangular, very high adradially, pm2 well developed, deep, inclined to ossicle width, pa4 prominent, elliptical, pa3 very prominent, subcircular, pm3 elongate, outline irregular; dm1 small, subtriangular, da2 prominent; outer surface large, with furrow row of spine bases along adradial margin, two or three rows of about four moderately large discrete spine bases on oral surface. Adamb comparisons: Distinguished by relative proportions and the broad, flat outer face with numerous prominent spine bases.

Pax. Ossicle small, tabula subspherical expansion of moderately long column, base small, subcircular.

Material.—One specimen, UCMP D-1944, one specimen, USNM 31523; both Pacific coast of North America; Recent; ossicle descriptions based on proximal, medial ossicles of one arm of UCMP specimen; R = 102 mm, r = 20 mm; hypotype no. 10671.

Genus *Thrissacanthias* Fisher, 1910
Thrissacanthias penicillatus (Fisher), 1905
(pl. 15, figs. 25–36)

Persephonaster penicillatus Fisher, 1905:297–298.
Thrissacanthias penicillatus: Fisher, 1911:79–85; pl. 17, fig. 4; pl. 18, figs. 1–5; pl. 53, fig. 1.

Description.—SupM (pl. 15, figs. 28–30). Ossicle massive, outer face outline of interbrachial ossicles quadrate, arm ossicles rectangular; outer face surface conical, with prominent circular spine base slightly oral of medial, forming surface apex; elongate, shallow concavities parallel distal, proximal margins; most ossicles with two additional moderately prominent medially located spine bases, usually one next to aboral margin, one slightly offset from margin; interbrachial ossicles with only one aboral spine base; bases very weakly developed on distal ossicles; remainder of surface covered by moderately densely spaced moderate-sized spine bases; side faces similar, subtriangular, without fasciolar surface, bordered abradially by well developed articulation ridge, moderately well developed ridge along adradial margin, no ridge along oral margin; intermarginal face gently concave, bordered adradially, abradially by low articulation ridges.

InfM (pl. 15, figs. 31–33). Ossicle massive, outer face outline subquadrate; outer face with two prominent U-shaped spine bases, open side of U generally directed slightly proximal of oral; aboral base usually larger, slightly proximal of oral base; oral margin irregular, somewhat arced; surface curvature, spinelet base development similar to that of SupM; side faces similar to those of SupM except abradial ridge may be incomplete orally; inner face with one or two prominent superambulacral bosses; intermarginal face similar to corresponding face of SupM, adradial articulation ridge very weakly developed. Marginal comparisons: Marginal series similar to each other, but distinct from those of other species considered in their massive convex form and prominent spine bases; the most significant difference between the two series within *Thrissacanthias* is in the nature of the development of spine bases.

Amb (pl. 15, figs. 34–36). Ossicle very fragile, low; amb body long, triangular; dentition well developed, consists of vertical medial plates, horizontal lateral plates; Unt G small, deep; oral apophyse small, massive; pGG short, triangular, dGG outline ovate, widest abradially; N low, pN on ridge, aboral ridge flat, weakly developed; amb body of disc ossicles proportionately short. Amb comparisons: Distinguished by the long, triangular amb body, low, rounded aboral ridge, and highly asymmetrical GG.

Adamb (pl. 15, figs. 25–27). Features similar in basic structure to *Platasterias* and *Luidia;* pa1 shallow, pm1 small, triangular; dm2 subcircular, centrally located, pa4 prominent, round, pa3 prominent, pm3 moderately large, subquadrate; oral surface large, flat, with two rows of about four prominent, moderate-sized, discrete spine bases, a third row of partially fused bases along the adradial margin; dm1 small, triangular. Adamb comparisons: Distinguished by relative proportions.

Pax. Ossicle low, tabula outline rounded, with spinelet bases, grades into column; two articulation projections oriented parallel to the arm axis.

Material.—One specimen, UCMP D-1955; one specimen, USNM E10429; both Pacific coast of North America; Recent; ossicle descriptions based on ossicles of UCMP specimen, R = 112 mm, r = 21 mm; hypotype no. 10672.

Genus *Dipsacaster* Alcock, 1893
Dipsacaster eximius Fisher, 1905
(pl. 15, figs. 1–11)

Dipsacaster eximius Fisher, 1905:296–297. Fisher, 1911:86–90; pl. 11, figs. 1, 2; pl. 13, fig. 2; pl. 14, fig. 1; pl. 16, fig. 3; pl. 52, fig. 1.

Description.—SupM (pl. 15, figs. 6–8). Ossicle massive, side face outline quadrate, irregular; outer face margins parallel, angular, one small spine base located medially on distal margin, remainder of surface covered by fine, densely spaced spinelet bases; fasciolar surfaces of side faces concave, sharply set off from triangular articulation surfaces, articulation ridges well developed adradially, abradially, continuous across aboral side face with corresponding ridges on opposite side face; articulation ridges discontinuous orally; aboral side face notched for paxilla. SupM comparisons: Distinguished by overall form.

InfM (pl. 15, figs. 9–11). Ossicle massive, outline quadrate; outer face outline in proximal ossicles tapers abradially; medial, distal ossicles with medial constriction; surface with several very slightly enlarged oral surface spine bases near distal margin, remainder of surface covered by small, densely spaced spinelet bases; side faces similar, outline rectangular except for aboral abradial prominence in proximal ossicles; fasciolar surface concave, with abradial fasciolar spinelet bases; articulation surface small, articulation ridge along border with fasciolar surface; aboral adradial edge inflated, intermarginal side face concave, smooth, with adradial, abradial articulation processes. InfM comparisons: Distinguished by overall form.

Amb (pl. 15, figs. 3–5). Ossicle low, body outline triangular; dentition prominent, consists of horizontal marginal plates, vertical medial plates; Unt G, oral groove well developed, deep; oral apophyse weak, dGG wide, rhomboidal, pGG relatively narrow; N elongate, low, pN on low ridge; aboral ridge weakly developed, rounded; amb body of disc ossicles proportionately short. Amb comparisons: Most similar to *T. penicillatus* Ambb, from which they differ in GG development.

Adamb (pl. 15, figs. 1, 2). Ossicle outline irregular; ad pm weakly developed, pa1 large, deep; pm1 wide, moderately high; pm2 elongate, elliptical; pa3, pa4 weakly developed, pm3 large, quadrate; dm1 moderate, triangular; outer face arched, with row of low, fused spine bases at peak of arch. Adamb comparisons: Distinguished by relative proportions, outer face development.

Pax. Ossicle very high, club-shaped, irregular spinelet bases on tabula, base very small.

Material.—One specimen, UCMP D-1555, numerous USNM specimens, 31950–31952, California; Recent; ossicle descriptions based on one arm of UCMP specimen, R = 110 mm, r = 45 mm; hypotype no. 10670.

Family BENTHOPECTINIDAE
Genus *Cheiraster* Studer, 1883
Cheiraster gazellae Studer, 1883
(pl. 16, figs. 1–15)

Cheiraster Gazella Studer, 1883:130–131.
Cheiraster gazellae: Fisher, 1919:196–200, pl. 50, figs. 1–3; pl. 51, fig. 1; pl. 52, fig. 1; pl. 54, fig. 3; pl. 56, figs. 1, 2.

Description.—SupM (pl. 16, figs. 1–4). Ossicle globose, outline of outer face polygonal, surface subspherical, with one large central spine base, remainder of surface covered by moderate-sized, moderately densely spaced spine bases; aboral margin of outer face arced, curves into inner face; aboral face narrow, inconspicuous groove; side faces variable, small, usually not parallel, ossicle tapers aborally, proximal side face frequently imbricate over distal side face of next proximal SupM, with weak articulation ridges along outer face margin, remainder of surface smooth; inner face smooth, some ossicles with one or two processes that contact Amb aboral ridge; intermarginal face generally two-faceted, larger distal facet slightly imbricate under InfM, smaller proximal facet slightly imbricate over InfM; with well developed articulation ridge along outer face, remainder of surface smooth; interbrachial ossicles proportionately shorter, usually with marginal, irregular, curved ridges of small bases, which form base of pedicellariae, inner face without amb contact processes.

InfM (pl. 16, figs. 5–8). Outline polygonal, asymmetrical; outer surface subspherical, one large medial spine base near aboral margin, two or three variable moderate-sized spine bases oral to large base, many ossicles with inconspicuous low semicircular pedicellaria base at oral margin, remainder of surface covered by small-to-moderate-sized moderately densely arranged spine bases; side faces similar, parallel, bordered abradially by low articulation ridge; oral side face narrow, inconspicuous, inner face irregular, smooth, with low processes for contact with Amb, Adamb; intermarginal face two-faceted, distal facet overlaps SupM, proximal facet overlapped by SupM, facets bordered by low articulation ridges, surfaces smooth, flat; interbrachial ossicles large, relatively short, not inclined, with relatively prominent side face articulation ridges, without inner face prominences. InfM comparisons: Most similar to SupMM of same species, from which they differ in spine base development and outline.

Amb (pl. 16, figs. 9–12). Ossicle long, high; amb body moderately long, subquadrate, dentition weakly developed, Unt G weak, oral apophyse strong, ridge sloped; dGG long, aboral outline triangular, overlaps short pGG of next distal Amb; pGG with shallow notch on aboral surface for articulation with dGG of next proximal Amb, dGG extends to form broad platform under amb extension; N subequal, prominent, extend laterally; aboral ridge very high, sharp, terminates abruptly with aboral prominence against marginal ossicles in arm; tapers abradially in disc ossicles. Amb comparisons: Distinguished by rectangular amb body.

Adamb (pl. 16, figs. 13–15). Ossicle outline square, ad pm prominent; outer surface with single, very prominent central spine base surrounded, except distally, by circle of small spine bases; furrow margin bordered by small bases; dml large, da1, da2 small, distinct, abradial margin beveled, articulates with InfM; proximal face largely occupied by smooth, irregular pm2; pa4 moderate, sharply defined, pm3 small, deep; ossicles of arm proportionately small, with proportionately smaller dml. Adamb comparisons: Distinguished by spine base pattern.

Pax. Small, simple, conical ossicle.

Material.—One specimen, CAS, Albatross Sta. 5538, Philippine Islands; Recent; ossicle descriptions based on proximal ossicles of one arm, R = 100 mm, r = 14 mm; hypotype UCMP no. 10673.

Genus *Pectinaster* Perrier, 1885
Pectinaster hylacanthus Fisher, 1913
(pl. 16, figs. 16–29)

Pectinaster hylacanthus Fisher, 1913:204–205. Fisher, 1919:187–190, pl. 48, fig. 2; pl. 54, fig. 1.

Description.—SupM (pl. 16, figs. 16–19). Ossicle small, globose; outer face outline variable, polygonal; outer face subspherical, with one large, central, upward-directed spine base surrounded by small, moderately densely arranged spine bases; outer faces curve into inner face, no distinct aboral side face; side faces similar, smooth, slightly concave; ossicles imbricate proximally; inner face smooth, with low Amb articulation prominences; intermarginal face bears two facets—smooth, concave, proximal facet slightly imbricate over InfM, distal facet slightly imbricate under InfM; interbrachial ossicles relatively smaller, shorter, more symmetrical, aboral outline trapezoidal; inner face without Amb contact prominence. SupM comparisons: Proximal ossicles very similar to those of *Cheiraster gazellae;* they differ in having smaller secondary spine bases, less distinct abradial articulation ridges on the intermarginal face and less prominent Amb contact prominences.

InfM (pl. 16, figs. 22–25). Outline polygonal, asymmetrical, outer face subspherical; with one large, subcentral spine base, another somewhat smaller base oral and distal to large one, moderate-sized moderately densely spaced bases extend orally from large bases to oral margin, other margins of face bordered by small moderately densely spaced spine bases; without distinct oral side face; side faces similar, subtriangular, concave, bordered abradially by low articulation ridge; inner face smooth, with low irregularities; intermarginal face bears two facets, bordered abradially by low articulation ridge, concave; interbrachial ossicles large, symmetrical, aboral outline trapezoidal, side faces strongly concave. InfM comparisons: Very similar to InfMM of *Cheiraster,* differ in the nature of the spine base development.

Amb (pl. 16, figs. 26–29). Ossicles long, high; amb body triangular, dentition fine; Unt G weakly developed, triangular, short, oral apophyse forms ridge; dGG broad, outline triangular, pGG small; N horizontal, aboral ridge prominent, sharp, with prominent aboral abradial projection in arm ossicles, projection lacking in disc ossicles. Amb comparisons: Ossicles lighter, with more distinctly triangular body than those of *Cheiraster.*

Adamb (pl. 16, figs. 20, 21). Ossicle polygonal, elongate, ad pm prominent; outer surface with one large spine base, smaller bases approximately circle it; dm1 small, narrow, da1, da2 moderately prominent; pm2 very prominent, pm3 small, pa3, pa4 sharply defined. Adamb comparisons: Very similar to *Cheiraster* Adambb, differ in being proportionately shorter, with weaker secondary spine bases.

Pax. Very small, flat, irregularly shaped ossicle.

Material.—7 specimens, CAS, Albatross Sta. 5445, Philippine Islands; Recent; ossicle descriptions based on proximal fragment of one arm of specimen of R = 65 mm, r = 7 mm; hypotype UCMP no. 10674.

Mistia spinosa, n. gen., n. sp.

(pl. 16, figs. 30–44; pl. 17, figs. 1–21, 35, 36)

Brisingid (?) Zullo et al., 1964:334.

Diagnosis.—Disc moderately large, with slender, tapering arms; margins and aboral surface heavily spinose; primary aboral ossicles large, conical, with single central base, surrounded by ring of smaller, generally inconspicuous bases; marginal ossicles inflated, area of spine bases not distinctly more inflated than remainder of ossicle, SupM with three principal spine bases, InfM with one principal spine base; Adambb with two transverse principal spine bases, relatively small ad pm. Mouth ossicle with three large adradial triangularly arranged spine bases, three moderate-sized spine bases abradial to large group, peristomial margin with low spine bases.

Description.—SupM (pl. 17, figs. 9–21). Ossicle massive, inflated, ovate, outer face subspherical; three large, medial, vertically aligned spine bases, ten to fifteen smaller spine bases loosely spaced, concentrated proximally and near larger bases; outer face overlaps distal flange of next proximal SupM, remainder of face smooth; side faces irregular smooth curved surfaces, distal face convex, imbricate under next distal SupM; no distinct aboral side face; inner face smooth, irregular; intermarginal face smooth, very irregular; medial ossicles proportionately lower, less distinctly imbricate; some secondary spine bases approach primaries in size.

InfM (pl. 16, figs. 38–43). Ossicle massive, inflated; outer face subspherical, one large subcentral spine base, about fifteen loosely spaced secondary spine bases approximately circling primary; remainder of surface smooth; outer face curves into inner face without a distinct oral side face; side faces irregular, smooth; inner face irregular, smooth; intermarginal face concave, smooth.

Amb (pl. 17, figs. 7, 8). Amb body long, outline rectangular, dentition fine; Unt G small, dGG articulation surface large; proximal ossicle with well developed dentition consisting of nearly horizontal plates; aboral ridge high, N horizontal, with prominent articulation structures.

Adamb (pl. 17, figs. 30–37). Ossicle massive, outline polygonal, ad pm moderately prominent; outer face large, with two large, transversely aligned spine bases, with small spine bases along adradial margin; abradial margin of arm ossicles beveled for contact with InfM (interactinal?) ossicles, dm1 outline triangular, da2 small, pm3 moderately well developed, pm1 very large.

Mouth ossicle. Ossicle massive, outer face outline triangular, peristomial margin lined with small spine bases, interossicle margin irregular, three large adradial spine bases triangularly arranged, three smaller transversely arranged spine bases abradial to large bases near oral margin.

Pax (pl. 17, figs. 1–6). Ossicle conical, sides convex, with one large spine base at apex, surrounded by very small bases, base outline subcircular to polygonal, polygon corners may extend as projections, oral surface inflated; arm ossicles rapidly decrease in diameter and elevation, distally on arm.

Proximal marginal spines (pl. 16, fig. 44; pl. 17, figs. 35–36). Conical, sharply pointed, cross section circular.

Aboral spines. Many incomplete spine fragments remain on the aboral surface; sizes are highly variable, ranging from an apparently complete spine just over 1 mm long to an incomplete spine over 13 mm long. These spines appear to have been less massive than the marginal spines.

Material.—One specimen, consisting of the disc and parts of four arms; Lower Oligocene Keasey Formation, near the town of Mist, Oregon; R of fragments 40 mm, 45 mm, 70 mm, 75 mm, r between 15 mm and 20 mm; UCMP A-5018; holotype no. 10675.

Discussion.—Because of preservation, it is difficult to determine whether aboral ossicles of varying sizes were intermingled, or whether ossicles were regularly distributed with respect to size; distribution of aboral ossicles on the surface suggests a general size gradation toward the arm tip, but some intergradation of ossicles size definitely occurred, as well. The smaller ossicles are more distinctly polygonal. The shape of the polygonal bases may be related to papulae distribution, the papulae fitting between the concave-sided polygons. If this is the function of the polygonal bases, there may have been no papular areas interbrachially, but the areas may have extended at least 20 mm or 25mm onto the arm.

The specimen is issigned to the Benthopectinidae on the basis of the relatively large marginals with long spines and the configuration of the marginals, Ambb, and Adambb. The conical primary aboral ossicles, with their regular polygonal bases, and the large, elongate, inflated marginal ossicles, with relatively few major spines, serve to separate *Mistia* from other benthopectinid genera.

Derivation of name.—*Mistia* (f.) commemorates the town of Mist, Oregon; *spinosa* (L., 'thorny') for the many large, prominent spines.

Family GONIASTERIDAE

Genus *Mediaster* Stimpson, 1857

Mediaster aequalis Stimpson, 1857

(pl. 18, figs. 26–37)

Mediaster aequalis Stimpson, 1857:530–531, pl. 23, figs. 7–11. Fisher, 1911:198–202; pl. 35, figs. 1–3; pl. 59, fig. 1.

Description.—SupM (pl. 18, figs. 32–34). Ossicle massive, outer face margins parallel, surface convex, margins rounded, surface covered by moderate-sized, densely spaced granule bases, some ossicles with small pedicellariae pits; side faces generally similar, bordered abradially by a broad, low, rounded, smooth articulation ridge that is continuous with' outer face; remainder of surface concave, bordered adradially by low, rounded ridge, side face depression larger on proximal side face; aboral side face in form of smooth, concave aboral notch; inner face smooth, slightly convex, intermarginal face smooth, concave, with articulation ridges along abradial, adradial margins; interbrachial ossicles similar, shorter, less massive with side faces subequal, abradial articulation ridges less prominent than on arm ossicles; distal ossicles similar, massive, aboral side face notch less strongly developed.

Amb (pl. 18, figs. 29–31). Ossicle massive, high, asymmetrical; amb body large, triangular; dentition weak, irregular, consists primarily of horizontal plates; Unt G, oral groove well developed, moderately deep, adradial apophyse massive, laterally faceted for Adamb articulation, abradial apophyse similar; GG similar, weakly developed; pN on strong proximal projecting ridge, dN with deep muscle depression, prominent articulation process; aboral ridge moderately well developed, rounded; amb body of disc ossicles proportionately short and high, aboral ridge with distinct adradial notch.

Adamb (pl. 18, figs. 26–28). Ossicle massive, outline quadrate; outer face rectangular, adradial margin notched, with three longitudinal rows of fused spine bases; side faces similar, dominated by large, smooth muscle depressions, with circular Amb ossicle articulation processes at adradial, abradial extremities of aboral margins; aboral surface with two short, transverse GG articulation grooves, proximal groove faint; adradial, abradial surfaces smooth, convex.

Aboral ossicles. Ossicle tabulate, tabula enlarged, circular or elongate, covered by small, circular, densely spaced granule bases, many with one polygonal pedicellaria; column short, massive, base low, with stellate articulation projections.

Material.—About twenty specimens, UCMP D-1944, California, (?); Recent; hypotype no. 10680.

Remarks.—In side view, the adradial portion of the ossicle appears extended into a prominent projection; this is a result of the aboral side face notch and the arced surface of the intermarginal face.

InfM (pl. 18, figs. 35–37). Oral side face large, smooth, concave, surface convex; ossicles otherwise similar to SupMM of the same species; distal ossicles of the two series not readily distinguished.

Genus *Paragonaster* Sladen, 1889

Paragonaster ctenipes hypacanthus Fisher, 1913

(pl. 18, figs. 14–25)

Paragonaster ctenipes hypacanthus Fisher, 1913:627. Fisher, 1919:228–232, pl. 70, fig. 3; pl. 71, fig. 2; pl. 72, fig. 1; pl. 91, fig. 9.

Description.—SupM (pl. 18, figs. 19–21). Ossicle massive; outer surface flat, covered with densely spaced granule bases, without large spine bases, face margins angular; side face bordered on all margins by low articulation ridge parallel to ossicle margins, central area of face slightly depressed, smooth, flat; SupM, InfM may be slightly offset, contact at offset in form of facet along oral side of proximal side face; aboral side face multifaceted (according to number of medial radial ossicles contacted), facets concave, smooth; inner face smooth, irregular; intermarginal face smooth, slightly concave; interbrachial SupM proportionately wider, shorter, cuneate.

InfM (pl. 18, figs. 23–25). Ossicle massive; outer face flat, with four or five very small, transversely arranged spine bases; SupM, InfM may be slightly offset, contact at offset in form of facet along aboral margin of distal side of face of InfM; no oral side face developed; inner face at right angles to intermarginal face; intermarginal face flat, featureless; interbrachial ossicle relatively short, wide; intermarginal face concave, at about 60° to inner face; number of spine bases greatest on proximal ossicles; other features similar to SupM.

Amb (pl. 18, figs. 14–16). Ossicles small, amb body subrectangular, very long; dentition very weak, Unt G shallow, without distinct oral groove, adradial apophyse weak; GG similar, outline subcircular; N low, triangular, on prominent ridges; aboral ridge moderately high, sharp.

Adamb (pl. 18, figs. 17, 18). Ossicle hook-shaped; outer face with one or two central spine bases encircled by ring of very small bases; adradial margin bordered by very small bases; M rectangular, aboral face smooth, irregular, articulation surfaces not strongly developed; adradial, abradial faces smooth.

Medial radial ossicle (pl. 18, fig. 22). Aboral outline polygonal, outer face flat with densely spaced granule bases, margins beveled, side faces may be multifaceted, smooth.

Material.—One specimen, CAS; 2 specimens, USNM 40168; all Recent; ossicles described from CAS specimen, R = 56 mm, r = 14 mm; hypotype UCMP no. 10679.

Genus *Ceramaster* Verrill, 1899
Ceramaster leptoceramus (Fisher), 1905
(pl. 18, figs. 1–13)

Tosia leptocerama Fisher, 1905:306–307.
Ceramaster leptoceramus: Fisher, 1911:210–214, pl. 39, figs. 1–3; pl. 58, fig. 3; pl. 60, fig. 2.

Description.—SupM (pl. 18, figs. 8–10). Ossicle massive, outline rectangular; outer surface very irregular, abradial portion covered by small, densely spaced granule bases, a raised area of irregular outline occurs adradially on outer surface, this area generally widest near ossicle margins, may be discontinuous near middle of face, may have scattered granule bases, numerous small, rectangular pedicellariae pits; face margins angular; side faces similar, irregular, smooth; intermarginal face smooth, flat abradial margin beveled, corner between side face and intermarginal face commonly beveled; aboral side face in form of concave, irregular groove; inner face irregular, gently convex.

InfM (pl. 18, figs. 11–13). Ossicle massive, similar in appearance to SupM; face margins angular, surface generally similar to that of SupM; side faces similar, irregular, smooth; oral side face an irregular groove; intermarginal face smooth, concave, bordered adradially, abradially by ridges. InfM comparisons: Most similar to SupMM of the same species, but raised areas not as distinct, fewer pedicellariae pits; intermarginal face of InfM with abradial ridge; distal InfMM are very similar to distal SupMM and cannot always be readily separated, distal InfMM are a bit wider than their corresponding SupMM and have a somewhat concave intermarginal face, as opposed to the convex intermarginal face of the SupMM.

Amb (pl. 18, figs. 4–7). Ossicle delicate, asymmetrical, extreme pN ridge development results in sinuous appearance; amb body wide, obtuse triangular outline; dentition fine, consists of irregular horizontal plates; Unt G well developed; GG small, subequal, outline semi-elliptical; pN on distinctly projecting ridge, dN on aboral surface; neither N strongly developed; aboral ridge low, rounded.

Adamb (pl. 18, figs. 1–3). Ossicle massive, rectangular, asymmetrical; oral surface notched adradially, furrow margin lined by fine fused spine bases, remainder of surface with irregular small spine bases, may have single adradial pedicellaria pit; side faces similar, M shallow, aboral margins of proximal and distal faces each have two small circular amb articulation processes, these processes may be contiguous on distal surface; aboral surface with two small, transversely elongate articulation depressions.

Aboral ossicle. Ossicle massive, tabula subcircular-to-rectangular, with small granulae depressions, may have pedicellariae pits; sides of column irregular, base with numerous short articulation projections.

Material.—One specimen, UCMP D-1949; numerous specimens USNM 31525; all from west coast of North America; Recent; ossicle descriptions based on arm of UCMP specimen, R = 60 mm; r = 36 mm; hypotype no. 10678.

Nehalemia delicata, n. gen., n. sp.
(pl. 17, figs. 37; pl. 18, figs. 38–48)

Diagnosis.—Flat starfish with long, slender, gently tapering arms and small subpentagonal disc; paxillae extend at least 5 mm out arm of base width of about 8 mm; SupM probably occupy most of aboral surface in remainder of arm; marginals distinctly encroach on disc, marginals small, short, with rounded margins, large granule bases, spine bases apparently lacking; paxillae small, columnal, tabulae not greatly enlarged.

Description.—Marginals (pl. 18, figs. 39–48). Ossicle massive; outer face evenly curved, gently convex, face margins rounded, surface apparently without spine bases, covered by moderately large, moderately densely spaced granule pits; side faces similar, flat, bordered along outer face margin, adradial margin by small articulation ridge; aboral, oral side faces rectangular, small, irregular; inner face rectangular, smooth, flat; intermarginal face flat or slightly concave, bordered by very low ridges; medial interbrachial ossicles proportionately wider, shorter; interbrachial ossicles from arm-disc junction cuneate, wide; medial, distal ossicles proportionately low, long.

Remarks.—Marginal ossicles of both series are essentially similar, though the inner face is commonly better developed on the SupM; this face is weak or lacking in proximal and interbrachial ossicles, the area being taken up by enlarged oral and aboral side faces.

Adamb (pl. 18, fig. 38). Hook-shaped ossicle, outer face with massive spine bases, ad pm well developed; proximal, distal side faces with prominent muscle depressions, dml on outer face. Adamb comparisons: Essentially the same morphology as *P. ctenipes,* ad pm weaker, spine bases probably stronger.

Pax (pl. 17, fig. 37; pl. 18, fig. 38). Ossicle short, massive, columnar, tabula massive.

Discussion.—The specimen is assigned to the Goniasteriade on the basis of general shape; the marginals and Ambb are typical of the family, the Adambb are very similar to those of *P. ctenipes,* but sharply set off from genera of other families considered. The subpentagonal shape, the long, slender arms, and the rounded, relatively small marginals distinguish *Nehalemia* from other goniasterid genera.

Derivation of name.—*Nehalemia* (f.) commemorates the Nehalem River of northwestern Oregon; *delicata* (L., 'delicate') refers to the delicate form of the species.

Material.—One leached specimen, consisting of the disc and proximal portions of three arms; the specimen apparently suffered some distortion and loss of ossicles prior to and during preservation, but little, if any, mechanical abrasion; R of 3 preserved arm fragments about 30 mm, 40 mm, and 40 mm, r about 12 mm from the lower Oligocene Keasey Formation near the town of Mist, northwestern Oregon; specimen is one of two undetermined asteroids recorded in Zullo et al., 1964; UCMP A-5018, holotype no. 10677.

Sucia suavis, n. gen., n. sp.
(pl. 19, figs. 1–16)

Diagnosis.—Stellate species with relatively large disc, sharp interbrachial angles, and rounded, sharply tapering arms; marginals covered by polygonal granules; paxilliform aboral ossicles on disc and corresponding to about the first five marginals from the center of the interbrachial arch; a single series of aboral ossicles present in the remainder of the arm.

Description (pl. 19, figs. 9, 10).—Marginals encroach well onto arms, but paxillar area appears to have at least almost reached arm tips; arm width at base about 13 mm, paxillar area width about 8 mm; arm width at 25 mm distal to base, 9 mm, paxillar area width, 2 mm; interbrachially, marginals extend little onto disc; marginal series correspond and are opposite each other, about five marginals in 10 mm proximally.

Marginals (pl. 19, figs. 2, 4-16).—SupMM short, high, wide; outer face outline semicircular, surface covered by moderately large polygonal granules; side faces similar, bordered abradially by low articulation ridges, very flat, with many shallow, regularly arranged depressions; aboral side face, inner face continuous, rectangular, flat, smooth; intermarginal face concave; InfMM generally similar to SupMM; narrow on arm, wide and deep on interbrachial arc.

Amb (pl. 19, fig. 16).—Narrow, massive ossicle.

Adamb (pl. 19, figs. 1, 10).—Arm ossicles hook-shaped, large outer face without distinctive spine bases, prominent ad pm, large dml; proximal Adambb larger, more massive, without prominent ad pm.

Pax (pl. 19, figs. 3, 9, 13, 16). Arm paxillae with large, quadrate bases, small summits.

Material.—The holotype is the only known specimen. It is from Sucia Island of the San Juan Islands, Washington State. The starfish is incomplete, consisting of most of the disc and three arms, plus a small portion of a fourth arm. The specimen is highly fragmentary, but apparently not seriously distorted. The specimen is partially enclosed in a highly indurated matrix. Preserved arm lengths are 44 mm, 53 mm, and 47 mm. Disc radius is approximately 17 mm. The longest arm fragment appears to be essentially complete; arm length in the living animal was probably only a few millimeters greater than in the fragment. Late Cretaceous, UCMP A-4534, holotype no. 10665.

Discussion.—The specimen is assigned to the Goniasteridae because of its general form and because of the prominent, rounded opposite marginal series. The morphology of the marginals, adambulacrals, and aboral ossicles is basically similar to such goniasterid genera as *Mediaster, Paragonaster,* and *Ceramaster.* However, gross morphology and morphology of discrete ossicles is distinct.

Derivation of name.—*Sucia* (f.) commemorates Sucia Island, northwestern Washington State; *suavis* (L., 'sweet') provides a play on words.

Goniasterid species A

(pl. 17, figs. 22–34)

Descriptions.—Marginals (pl. 17, figs. 28–34). Ossicle massive, rectangular, outer face outline rectangular, surface flat, covered by large, moderately densely spaced granule bases; margins angular; side faces similar, deeply concave, bordered abradially by articulation ridges; inner faces rectangular, flat or concave, smooth, marginals of the two series are generally similar.

Amb (pl. 17, figs. 22, 23).—Ossicle massive; amb body very low, very long, triangular, amb extension with broad, rounded medial hump; dentition moderately well developed, primarily vertical plates; Unt G well developed, oral groove weakly developed; apophyses moderately well developed; GG narrow, massive; N low, elongate, similar, pN on prominent ridge.

Adamb (pl. 17, figs. 24–27). Ossicle high, massive; outer face flat, with three longitudinal ridges; adradial face concave, ad pm prominent; proximal, distal faces smooth, concave, aboral face with L-shaped Amb articulation facets.

Pax. Ossicle massive, columnar, tabula small, stellate, base stellate.

Material.—Fragment of disc and about 30 mm of one arm of one specimen, from the Lower Oligocene Keasey Formation near the town of Mist, northwestern Oregon. Specimen is one of two unassigned asteroids mentioned in Zullo et al., 1964; UCMP A-5018; hypotype no. 10676.

Discussion.—Not enough of the specimen is preserved to allow accurate determination of gross shape, but the specimen appears to have been moderately large. Because relatively small marginals and Ambs are present near to larger ossicles of the disc, it appears probable that the specimen had relatively short, rapidly tapering arms, similar to some species of *Ceramaster.* The proportionately small marginal ossicles, their flat surface regularly covered by moderately large granule depressions, and the massive, stellate paxillae serve to distinguish this species from other goniasterids considered here. Morphology of Recent genera of goniasterids is not well enough known, however, to warrant erection of a new taxon.

SUMMARY

In this study, the utility and limitations of discrete ossicles in taxonomic and evolutionary interpretations are evaluated for certain fossil and Recent asteroids. A number of Recent starfish have been disarticulated and their ossicles compared and described in detail. Necessary new terminology is proposed (see glossary).

Ossicle morphology has been found to be a very useful tool for use in species discrimination, even among relatively closely related species, such as those belonging to the genus *Luidia*. Interspecific variation appears greater than intraspecific variation in the species studied.

A number of limitations must be considered, however, in the use of ossicle morphology. There are significant intraspecific morphological differences related to ontogeny. Ossicles from distal areas of the arm tend to be more massive and possess simpler morphology than the large proximal ossicles. Distal ossicles from large specimens tend to be more massive than proximal ossicles of similar dimensions from smaller specimens. Great variation in ossicle morphology may be associated with certain positions on the starfish. For example, the closely spaced near-mouth adambulacrals and ambulacrals of certain species appear light and relatively strongly overlapping when compared to ossicles from farther out on the arm. Variation of these types is repeated in various taxa and is therefore predictable.

As in any group of organisms, discrimination is aided by more information; all species considered here can be recognized by studying multiple ossicle types, but not all are distinctive if only one or two ossicle types are considered.

In addition to being useful in species discrimination, ossicle morphology also provides information important for the grouping of taxa above the species level and therefore, presumably, for the recognition of evolutionary trends. For example, the "groups" of Döderlein in the genus *Luidia* are reflected in ossicle morphology; the genus as a whole shows considerable unity of ossicle morphology. The ambulacral-adambulacral articulation structures of the considered species of the Astropectinidae are quite similar to one another, but quite distinctive from that found in the Benthopectinidae, a family whose members also have a unified, distinctive pattern of ambulacral-adambulacral articulation. *Archaster,* a genus of uncertain affinities, has some relatively distinctive features of ossicle morphology.

Recognition of higher taxa and evolutionary trends is aided by the apparently differing evolutionary rates in different ossicle types. In general, marginal ossicles were found to be diverse in form, with considerable variation occurring among genera of a single family. Perhaps because of their complex system of muscle attachment and articulation processes, ambulacral and adambulacral ossicles were found to be much less variable and therefore, presumably, more conservative in evolution. Differing evolutionary rates may also have occurred on different parts of ossicles. For example, among the adambulacral ossicles of studied species of the Luidiidae, Astropectinidae, Benthopectinidae, and *Archaster,* there is more similarity among interadambulacral articulation structures than there is among adambulacral-ambulacral articulation structures.

On the ambulacrals and adambulacrals, it is possible to recognize apparently homologous structures in the same relative positions performing similar functions in taxonomically diverse species. A broad range of expression of these features does exist; an intensive study of these structures from diverse taxa, therefore, will provide new data on relationships within the subclass.

Enough morphological unity exists within a given ossicle type between various higher taxa to suggest that careful comparisons of discrete ossicle morphology in many starfish may provide information useful in the determination of homologous ossicle types. Here, the superomarginals of the Asteropectinidae are considered probably to be homologous to the first paxillae of the Luidiidae, although the similarity of position and function may have led to a misleading convergence of form.

Ossicle morphology is used as the basis for change in assignment of two taxa above the species level. *Luidia foliolata* Grube was assigned to the Clathrata Group by Döderlein; its ossicle morphology is typical of the Alternata Group in every important aspect, so it is considered here to be a member of the Alternata Group. *Luidia (Platasterias) latiradiata* Gray, recently considered to be a living somasteroid, has ossicle morphology very similar to that displayed by the Clathrata Group of *Luidia.* The morphological differences between the Ciliaris and Clathrata Groups were found to be greater than the differences between *Platasterias* and the Clathrata Group. *Platasterias,* therefore, is not considered here to be a living somasteroid; rather, it is placed in the Clathrata Group of the genus *Luidia.* Once ossicle morphology has been described for more species of *Luidia,* it may prove desirable to recognize the Ciliaris Group as a distinct genus.

Certain aspects of overall asteroid form may be reflected in the morphology of individual ossicles. It is therefore possible for a paleontologist to partially reconstruct a starfish from discrete ossicles. For example, the acute interbrachial angles of *Luidia* are reflected in the very short, cuneate, interbrachial marginals. The narrow arms are reflected in the presence of a superambulacral boss on the inferomarginals; the arms had to be narrow enough to permit the superambulacral ossicle to extend from the ambulacral to the inferomarginal.

The distinctive nature of ossicle morphology has allowed recognition of a number of new taxa of fossil west American starfish. *Luidia etchegoinensis,* from Pliocene rocks, and *L. sanjoaquinensis,* from Pleistocene rocks, are new species of the Alternata Group of *Luidia. L. foliolata,* abundant in the living west American faunas, is recorded from Pliocene or Pleistocene rocks. Specimens of *Luidia* are also reported from three other localities from Pliocene and younger rocks. Poor preservation precludes assignment of these specimens to species. The benthopectinid *Mistia spinosus* and the goniasterids *Nehalemia delicata* and *Sucia suavis* are new genera and species described from Cretaceous and Oligocene rocks. A species of *Astropecten* and a goniasterid species are reported from Oligocene rocks but are not named because of inadequate knowledge of ossicle morphology of modern members of these taxa.

The small sample of fossil taxa studied here provides no evidence of major change in the nature of the west American asteroid fauna during the Cenozoic.

LOCALITY DESCRIPTIONS

Locality UCMP A-415
Pico Formation, Pliocene; Santa Paula Quad., 1942, 1:62500.
South flank of Sulfur Mountain, at the foot of the steep slope between Santa Paula Canyon and Coche Canyon; about 4 miles southeast of Ojai, Calif.

Locality UCMP A-1461
"Etchegoin" Formation, Pliocene; Nipomo Quad., 1952, 1:62500, T32S R15E.
On the north side of Long Canyon at an elevation of about 1,150 feet, in the head of a small canyon about 4,300 feet N85E of BM 922, near Park Ranch; about 9 miles northeast of Nipomo, Calif., twenty miles southeast of San Luis Obispo, Calif.

Locality UCMP A-4534
Nanaimo Group, Upper Cretaceous; Orcas Island Quad., 1943, 1:62500.
The tip of the southern margin of the cove northwest of Fossil Bay, Sucia Island; about 3 miles north of Orcas Island, San Juan County, Wash.

Locality UCMP A-5018
Keasey Formation, Lower Oligocene; Keasey Quad., 1943, 1:62500.
Shale cliffs exposed for about 200 yards along the west side of the Nehalem River nearly ¼ mile south of the secondary highway from the town of Mist; about 35 miles northwest of Portland, Ore.

Locality UCMP B-5046
Etchegoin Formation, Pliocene; La Cima Quad., 1934, 1:31680, midpoint of the west boundary of Sec. 16, T22S R18E.
Along the north bank and near the head of Arroyo Doblegado, just below the midpoint of the west boundary of Sec. 16. Locality includes about 35 feet of light brown, silty, medium-to-fine-grained, friable sandstone found on the north bank of the dry creek. The arroyo strikes about N75W, and collecting extends about 0.3 mile to the east; Kettleman Hills, east of Avenal, Calif.

Locality UCMP B-7085
Temblor Formation, Miocene; Joaquin Rocks Quad., 1943, 1:62500, SW ¼ NW ¼ Sec. 21 R15E T19S.
An east-west trail enters the section from the west and terminates against a north-south trail in the NW ¼; locality is south of the junction on the east side of the north-south trail along the south wall of a pass in the Temblor Formation; fossils are from resistant ledges near the hilltop in a sand-dollar-rich "button-bed"; about 8 miles north of Coalinga, Calif.

Locality D-195
"Merced"Formation, Pliocene; Ano Nuevo Quad., 1943; 1:62500.
The Pacific Ocean beach about 1 mile northwest of the Santa Cruz-San Mateo County line near Ano Nuevo Point in San Mateo County, Calif.

Locality UCMP D-2439
San Joaquin Formation, Pliocene; La Cima Quad., 1963, 1:24000, NW ¼ SW ¼ NW ¼ SE ¼ Sec. 6, T22S R18E.
On the north side of the arroyo about 300 feet downstream of its confluence with another arroyo; about 20 feet below the crest of the main ridge, which extends to the northwest; Kings Co., Calif.

Locality U. S. Geological Survey M-3917
Branch Canyon Formation, Miocene; New Cuyama Quad., 1964, 1:24000, T9N R27W.
In the echinoid biostromal beds on the ridge crest near the top of the formation, at an elevation of about 2,920 feet; 800 feet south and 1,700 feet west of the northeast corner of projected Sec. 12; Santa Barbara Co., Calif.

LITERATURE CITED

AGASSIZ, ALEXANDER
  1877. North American starfishes. Mem. Mus. Comp. Zool., 5:1–136.

BLAKE, D. BRYAN
  1972. Sea star *Platasterias:* ossicle morphology and taxonomic position, Science 176:306–307.

CLARK, AILSA M.
  1953. Notes on asteroids in the British Museum (Natural History): 3, *Luidia;* 4, *Tosia* and *Pentagonaster.* Bull. Brit. Mus. (Nat. Hist.) Zool., 1:379–412.

CLARK, HUBERT L.
  1910. The echinoderms of Peru. Bull. Mus. Comp. Zool., 52(17):321–358.

DÖDERLEIN, LUDWIG
  1917. Die Asteriden der Siboga-Expedition I: Die Gattung *Astropecten* und ihre Stammesgeschichte. Siboga-Expeditie Mono., 46a:1–192.
  1920. Die Asteriden der Siboga-Expedition II: Die Gattung *Luidia* und ihre Stammesgeschichte. Siboga-Expeditie Mono., 46b:193–294.

FELL, H. BARRACLOUGH
  1962. A surviving somasteroid from the eastern Pacific Ocean, Science 136:633–636.
  1963. The phylogeny of sea-stars. Phil. Trans. Roy. Soc. London, ser. B., 246:381–435.

FISHER, WALTER K.
  1905. New starfishes from deep water off California and Alaska. Bull. Bur. Fisheries for 1904, 24:291–320.
  1906. The starfishes of the Hawaiian Islands. Bull. U. S. Fish. Comm. for 1903, 3:987–1130.
  1911. Asteroidea of the North Pacific, 1: Phanerzonia and Spinulosa. Bull. U. S. Nat. Mus., 76(1):1–420.
  1913. New starfishes from the Philippine Islands, Celebes, and the Moluccas. Proc. U. S. Nat. Mus., 46:201–224.
  1919. Starfishes of the Philippine Seas and adjacent waters. Bull. U. S. Nat. Mus., 100(3): 1–712.

FORBES, EDWARD
  1848. On the Asteridae found fossil in British strata. Geol. Sur. Great Britain, Mem. 2:457–482.

GARCIA, E. R., and H. H. CAMACHO
  1965. Noticia sobre el hallazgo de restos de equinodermos en el paleoceno de General Roca (Pcia. Rio Negro). Ameghiniana 4:84–100.

GRAY, JOHN EDWARD
  1840. A synopsis of the genera and species of the Class Hypostoma (*Asterias* Linn.). Ann. Mag. Nat. Hist., ser. 1. 6:175–184, 275–290.
  1871. Description of *Platasterias,* a new genus of Astropectinidae, from Mexico. Proc. Zool. Soc. London, pp. 136–137.

GRUBE, A. E.
  1866. Einige neue Seesterne des hiesigen zoologischen Museums. 43 Jahresber. schlesischen Ges. väterländische Cultur, for 1865, pp. 59–61.

HEDDLE, DUNCAN
  1967. Versatility of movement and the origin of the asteroids. *in* N. Millot, ed., Echinoderm Biology. Zool. Soc. London no. 20, Academic Press, pp. 125–142.

HESS, HANS
  1955. Die Fossilen Astropectiniden (Asteroidea) Neue Boebachtungen und Ubersicht über die Bekannten Arten. Schweiz. Paläontol. Abhandl., 71:1–113.

HOWE, HENRY V.
  1942. Neglected Gulf Coast Tertiary microfossils. Bull. Am. Assoc. Pet. Geol., 26:1188–1199.

KESLING, R. V.
  1969. Three Permian starfish from Western Australia and their bearing on revision of the Asteroidea. Univ. Michigan, Mus. Paleontol. Contrib., 22:361–376.

LAMARCK, J. P. B. A. DE
  1816. Histoire naturelle des animaux sans vertebres, vol. 2.
LEHMANN, W. M.
  1957. Die Asterozoen in den Dachschiefen des rheinischen Unterdevons. Hess. Landesamt Bodenforsch., Abhandl., 21:1–160.
LORIOL, P. DE
  1891. Notes pour servir à l'étude des Echinoderms. Mem. Soc. Phys. Hist. Nat. Genève, supp. vol. 1890, no. 8.
LÜTKEN, CH.
  1859. Bidrag til Kundskab om de ved Kysterne af Mellem-og Syd-Amerika levende Arter af Söstjerner. Vidensk. Medd. naturh. Foren. Kbh., pp. 25–96.
MADSEN, F. JENSENIUS
  1966. The Recent sea-star *Platasterias* and the fossil Somasteroidea. Nature, 209:1367.
MARTENS, E. VON
  1865. Über ostasiatische Echinodermen. Arch. Naturgesch., 31:345–360.
MORTENSEN, TH.
  1925. Papers from Dr. Th. Mortensen's Pacific expedition 1914–16: 29, echinoderms of New Zealand and the Auckland Campbell Islands; 3–5, Asteroidea, Holothurioidea and Crinoidea. Copenhagen: Viden Skabelige meddelelser Dansk naturhistorisk forening, 79:261–419.
MÜLLER, ARNO HERMANN
  1953. Die isolierten Skelettelemente der Asteroidea (Asterozoa) aus der Obersenonen Schreibkreide von Rugen. Beih. Zeit. Geol., 8:1–66.
MÜLLER, J., and F. H. TROSCHEL
  1842. System der Asteriden. Braunschweig, 134 pp.
MURAKAMI, SHIRO
  1963. The dental and oral plates of the Ophiuroidea. Roy. Soc. New Zealand, Trans. Zool., 4:1–48.
NIELSEN, K. BRÜNNICH
  1943. The asteroids of the Senonian and Danian deposits of Denmark. Copenhagen: K. Danske videnskabernes selskab, Biologiske skrifter, 2:1–68.
PERRIER, EDMOND
  1876. Revision de la collection de stellérides du Muséum d'Histoire naturelle de Paris. Arch. Zool. Exp., 5:1–104, 209–304.
  1892. Sur la morphologie du squelette des étoiles de mer. Acad. Sci. Paris., Compt. Rend., 115:670–673.
PHILIP, G. M.
  1965. Ancestry of sea-stars. Nature, 208:766–768.
PHILIPPI, R. A.
  1837. Ueber die mit *Asterias aurantiaca* verwandten und verwechselten asterien der Sicilianischen Kuste. Arch. Natur, 3(1)193–194.
RASMUSSEN, H. W.
  1950. Cretaceous Asteroidea and Ophiuroidea, with special reference to the species found in Denmark. Denmarks Geol. Unders. ser. 2. 7:1–134.
  1965. The Danian affinities of the Tuffeau de Ciply in Belgium and the "Post-Maastrichtian" in the Netherlands. Mededelingen van de Geologische Stichting, n.s., 17:33–40.
RÉAUMUR, R. A.
  1732. Observatio de stellis marinis. Hist. Acad. Sci. Paris, vol. for 1710, nouv. ed., rev., cor., augm., pp. 439–490.
SAY, THOMAS
  1825. On the species of the Linnaean genus *Asterias* inhabiting the coast of the United States. Philadephia: Acad. Natur. Sci., 5(1)141–154.
SCHÄFER, WILHELM
  1962. Aktuo-Paläontologie nach Studien in der Nordsee. Frankfurt am Main: Verlag Waldemar Kramer, 666 pp.

SCHUCHERT, CHARLES
  1915. Revision of Paleozoic Stelleroidea with special reference to North American Asteroidea. Bull. U. S. Nat. Mus., 88:1–311.
SLADEN, WILLIAM PERCEY
  1880. Traces of ancestral relations in the structure of the Asteroidea. Proc. York. Geol. Poly. Soc. n.s., 7:275–284.
SLADEN, W. P., and W. K. SPENCER
  1891–1908. British fossil Echinodermata from the Cretaceous formations; vol. 2, Asteroidea and Ophiuroidea. Paleont. Soc. London, mon., pp. 1–138.
SPENCER, W. K.
  1914–1940. British Palaeozoic Asteroza. Paleontograph. Soc. London, mon., 540 pp.
  1951. Early Palaeozoic starfish. Phil. Trans. Roy. Soc. London, ser. B, 235:87–129.
SPENCER, W. K., and C. W. WRIGHT
  1966. Asterozoa, *in* R. C. Moore, ed. Treatise on Invertebrate Paleontology. Geol. Soc. Amer. and Kan. Univ. Press, pp. U4–U107.
STIMPSON, WILLIAM
  1857. On the crustacea and echinodermata of the Pacific shores of North America. Boston Nat. America. Boston Nat. Hist. Jour., 6:527–531.
STUDER, T.
  1883. Über die auf der Expedition S.M.S. Gazelle gesammelten Asteriden. Berlin: Sitzungsber. Gesellsch. Naturforsch. Freunde for 1883, no. 8:128–132.
VERRILL, A. E.
  1899. Revision of certain genera and species of starfish with descriptions of new forms. Trans. Conn. Acad. Arts Sci., 10:145–234.
  1914. Monograph of the shallow-water starfishes of the North Pacific coast from the Arctic Ocean to California. Smith. Inst. Harriman Alaska Series 10:1–408.
  1915. Report on the starfishes of the West Indies, Florida and Brazil. Iowa Univ. Bull. Lab. Natur. Hist., 7:1–232.
VIGUIER, C.
  1878. Anatomie comparée du squelette des Stellerides. Arch. Zool. Exper. Gen., 7:33–250.
WEBER, JON N.
  1968. Fractionation of the stable isotopes of carbon and oxygen in calcareous marine invertebrates—the Asteroidea, Ophiuroidea and Crinoidea. Geochimica et Cosmochimica Acta., 32:33–70.
ZULLO, V. A., R. F. KAAR, J. WYATT DURHAM, and E. C. ALLISON
  1964. The echinoid genus *Salenia* in the eastern Pacific. Paleontology, 7:331–349.

PLATES

NOTE ON TYPE NUMBERS IN PLATE DESCRIPTIONS

Unless otherwise stated, all numbers refer to the University of California Museum of Paleontology.

A separate number has been applied to each specimen and a letter to each illustrated ossicle removed from the specimen. Discrete ossicles of certain fossils have been illustrated which were not removed from the specimen; no separate letter designation has been applied to them.

PLATE 1

Figs. 1–27. *Luidia columbia* (Gray).   Hypotype no. 10637.

1–3, hypotype no. 10637a, right Adamb, oblique aboral, oblique proximal and oblique distal views.

4, hypotype no. 10637b, right Adamb, oblique proximal view.

5, 6, hypotype no. 10637c, left Adamb, oblique proximal and oblique distal views. Arrow points to oral flange.

7–9, hypotype no. 10637d, left Amd, proximal, aboral and oral views.

10–12, hypotype no. 10637e, right Amb, proximal, aboral and oral views.

13–15, hypotype no. 10637f, left distal InfM, oral, distal and proximal views.

16–18, hypotype no. 10637g, left medial InfM, oral, proximal and distal views.

19–21, hypotype no. 10637h, right medial InfM, oral, proximal and distal views.

22–24, hypotype no. 10637i, left proximal InfM, oral, proximal and distal views.

25–27, hypotype no. 10637j, right Amb, distal, aboral and oral views.

Figs. 28–53. *Luida* (*Platasterias*) *latiradiata* (Gray).   Hypotype 10634.

28, 29, hypotype no. 10634a, Pax, lateral and aboral views.

30, 31, hypotype no. 10634b, Fst Pax, lateral and aboral views.

32–34, hypotype no. 10634c, left distal Adamb, oblique aboral, oblique proximal and oblique distal views.

35–37, hypotype no. 10634d, left medial Adamb, oblique aboral, oblique proximal and oblique distal views.

38–40, hypotype no. 10634e, right Amb, lateral, aboral and oral views.

41–43, hypotype no. 10634f, left Amb, lateral, aboral and oral views.

44, 45, hypotype no. 10634g, left distal InfM, proximal and distal views.

46, 47, hypotype no. 10634h, right proximal InfM, distal and proximal views.

48–50, hypotype no. 10634i, left proximal InfM, oral, distal and proximal views.

51–53, hypotype no. 10634j, right proximal InfM, oral, distal and proximal views.

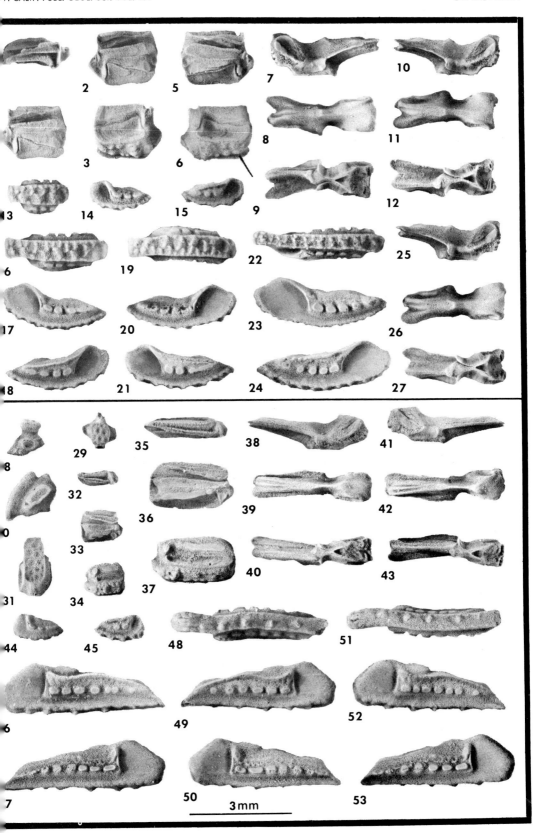

3mm

PLATE 2
Figs. 1–31. *Luidia senegalensis* (Lamarck).   Hypotype no. 10635.

1, 2, hypotype no. 10635a, Pax, lateral and aboral views.

3, hypotype no. 10635b, Fst Pax, lateral view.

4, 8, hypotype no. 10635c, left medial Adamb, oblique distal and oblique proximal views.

5, 9, hypotype no. 10635d, right medial Adamb, oblique distal and oblique proximal views.

6, 7, 10, hypotype no. 10635e, right proximal Adamb, oblique distal, oblique aboral and oblique proximal views.

11, 15, 19, hypotype no. 10635f, left distal InfM, oral, distal and proximal views.

12, 16, 20, hypotype no. 10635g, right medial InfM, oral, distal and proximal views.

13, 17, 21, hypotype no. 10635h, left medial InfM, oral, distal and proximal views.

14, 18, 22, hypotype no. 10635i, right proximal InfM, oral, distal and proximal views.

23, 26, 29, hypotype no. 10635j, right medial Amb, lateral, aboral and oral views.

24, 27, 30, hypotype no. 10635k, right medial Amb, lateral, aboral and oral views.

25, 28, 31, hypotype no. 10635l, left proximal Amb, lateral, aboral and oral views.

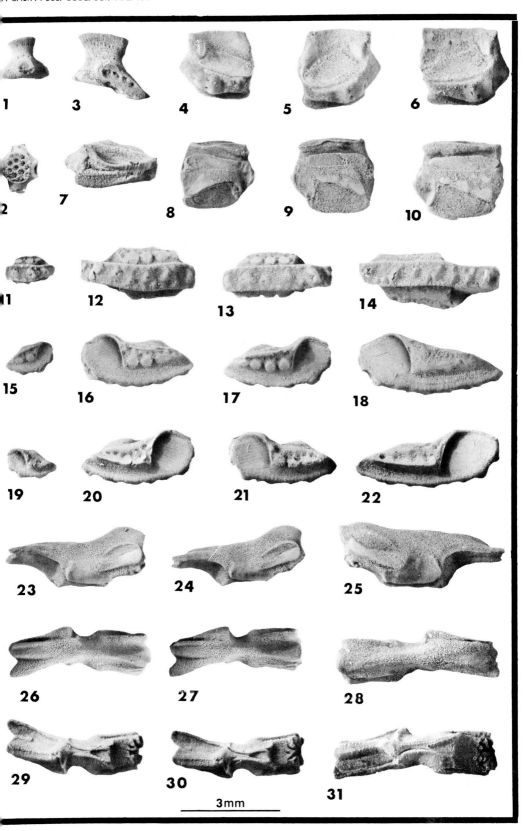

PLATE 3

Figs. 1–28. *Luidia clathrata* (Say).   Hypotype no. 10636.

1, 3, hypotype no. 10636a, Pax, aboral and lateral views.

2, 4, hypotype no. 10636b, Fst Pax, aboral and lateral views.

5, 6, 9, hypotype no. 10636c, left Adamb, oblique oral, oblique proximal and oblique distal views.

7, 8, hypotype no. 10636d, right Adamb, oblique proximal and oblique aboral views.

10, hypotype no. 10636e, left Adamb, oblique distal view.

11, 14, 17, hypotype no. 10636f, left Amb, lateral, aboral and oral views.

12, 15, 18, hypotype no. 10636g, left Amb, lateral, aboral and oral views.

13, 16, 19, hypotype no. 10636h, left Amb, lateral, aboral and oral views.

20, 23, 26, hypotype no. 10636i, right InfM, proximal, distal and oral views.

21, 24, 27, hypotype no. 10636j, right InfM, proximal, distal and oral views.

22, 25, 28, hypotype no. 10636k, left InfM, proximal, distal and oral views.

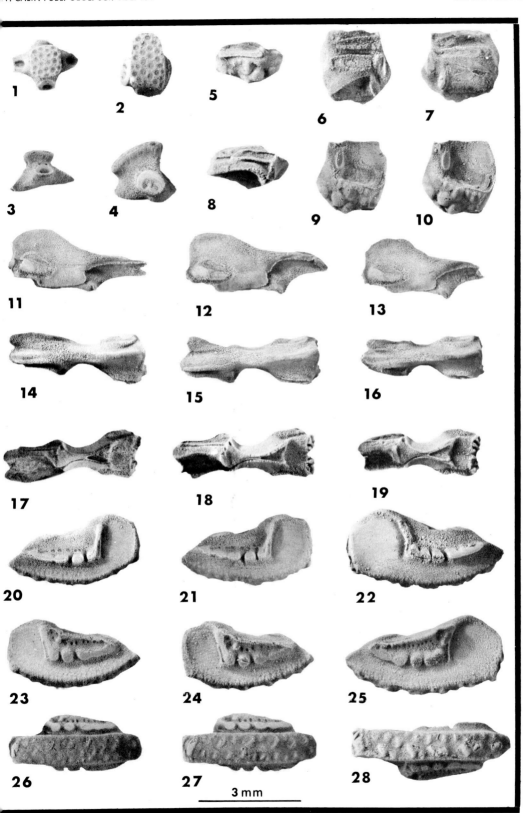

3 mm

PLATE 4

Figs. 1–26. *Luidia tesselata* Lütken.   Hypotype no. 10638.

1, 2, hypotype no. 10638a, Pax, aboral and lateral views.

3, 4, hypotype no. 10638b, Fst Pax, lateral and aboral views.

5–7, hypotype no. 10638c, left proximal InfM, distal, proximal and oral views.

8–10, hypotype no. 10638d, left medial InfM, distal, proximal and oral views.

11–13, hypotype no. 10638e, right medial InfM, distal, proximal, and oral views.

14, 15, hypotype no. 10638f, right Adamb, oblique proximal and oblique distal views.

16, 17, hypotype no. 10638g, left Adamb, oblique proximal and oblique distal views.

18–20, hypotype no. 10638h, right Adamb, oblique distal, oblique proximal and oblique aboral views.

21–23, hypotype no. 10638i, right Amb, lateral, aboral and oral views.

24–26, hypotype no. 10638j, left Amb, lateral, aboral and oral views.

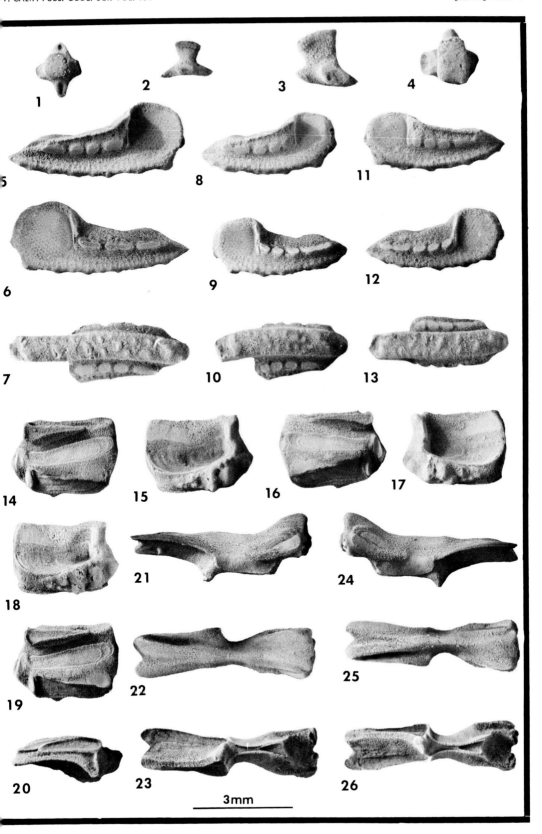

3mm

PLATE 5

Figs. 1–17. *Luidia ludwigi* Fisher. Hypotype no. 10640.

1, 2, hypotype no. 10640a, right Adamb, oblique proximal and oblique distal views.

3–5, hypotype no. 10640b, left Adamb, oblique proximal, oblique distal and oblique aboral views.

6–8, hypotype no. 10640c, right Amb, aboral, oral and lateral views.

9, 16, 17, hypotype no. 10640d, right Amb, lateral, oral and aboral views.

10, 11, hypotype no. 10640e, left InfM, proximal and distal views.

12, hypotype no. 10640f, right InfM, oral view.

13–15, hypotype no. 10640g, left InfM, proximal, distal and oral views.

Figs. 18–34. *Luidia magnifica* Fisher. Hypotype no. 10639.

18, hypotype no. 10639a, Pax, lateral view.

19–21, hypotype no. 10639b, right Adamb, oblique aboral, oblique distal and oblique proximal views.

22, hypotype no. 10639c, left Adamb, oblique proximal view.

23–25, hypotype no. 10639d, left Amb, aboral, oral and lateral views.

26–28, hypotype no. 10639e, left InfM, oral, distal and proximal views.

29–31, hypotype no. 10639f, left InfM, oral, distal and proximal views.

32–34, hypotype no. 10639g, left Amb, aboral, oral and lateral views.

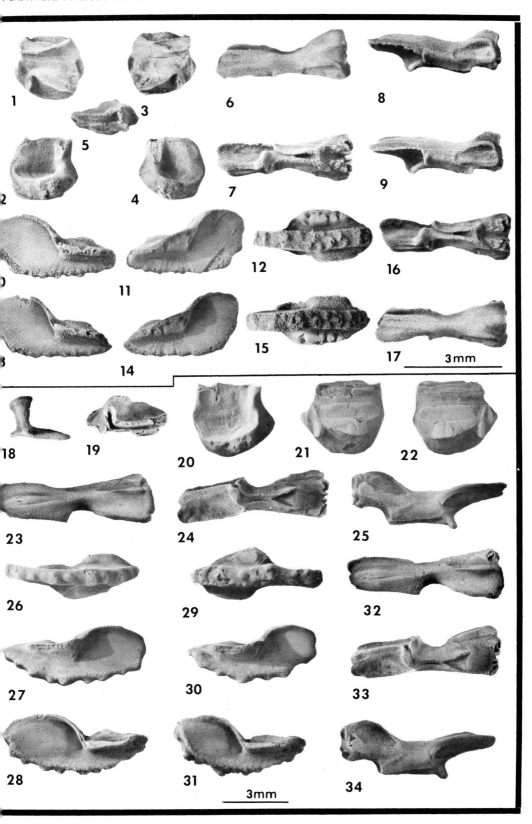

3mm

3mm

PLATE 6

Figs. 1–24. *Luidia mascarena* Döderlein.   Hypotype no. 10641.

1, 2, hypotype no. 10641a, Pax, lateral and aboral views.

3–5, hypotype no. 10641e, right Amb, lateral, oral and aboral views.

6–8, hypotype no. 10641f, right Amb, lateral, oral and aboral views.

9–11, hypotype no. 10641b, left Adamb, oblique aboral, oblique proximal, and oblique distal views.

12, 13, hypotype no. 10641c, right Adamb, oblique proximal and oblique distal views.

14, 15, hypotype no. 10641d, left Adamb, oblique proximal and oblique distal views.

16–18, hypotype no. 10641g, right InfM, oral, proximal and distal views.

19–21, hypotype no. 10641h, left InfM, oral, proximal and distal views.

22–24, hypotype no. 10641i, left InfM, oral, proximal and distal views.

Figs. 25–42. *Luidia alternata* (Say).   Hypotype no. 10642.

25, hypotype no. 10642a, left Adamb, oblique aboral view.

26, 27, hypotype no. 10642b, right Adamb, oblique proximal, and oblique distal views.

28–30, hypotype no. 10642c, right medial InfM, proximal, distal and oral views.

31–33, hypotype no. 10642d, left proximal InfM, proximal, distal and oral views.

34–36, hypotype no. 10642e, right interbrachial InfM, proximal, distal and oral views.

37–39, hypotype no. 10642f, left Amb, lateral, aboral and oral views.

40–42, hypotype no. 10642g, right Amb, lateral, aboral and oral views.

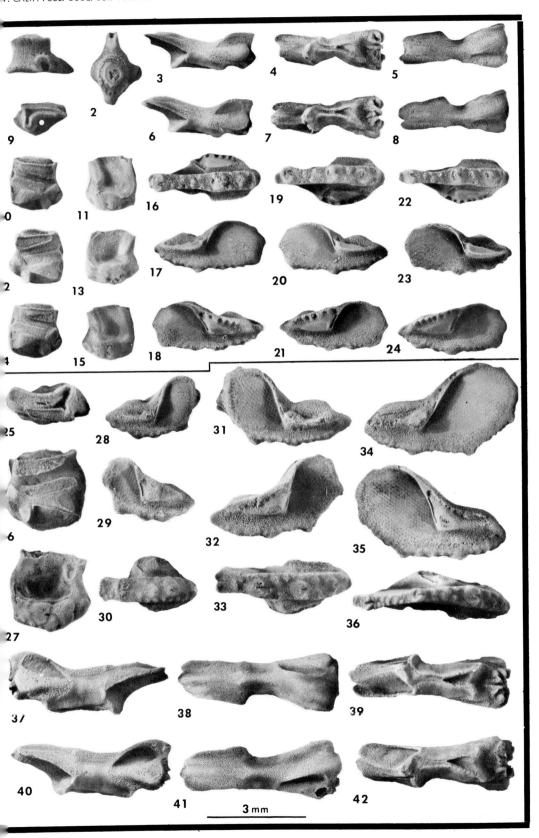

3 mm

PLATE 7

Figs. 1–19. *Luidia quinaria* v. Martens.   Hypotype no. 10643.

1, 2, hypotype no. 10643a, Pax, aboral and lateral views.

3–5, hypotype no. 10643b, left Adamb, oblique aboral, oblique distal and oblique proximal views.

6, 7, hypotype no. 10643c, right Adamb, oblique distal and oblique proximal views.

8–10, hypotype no. 10643d, left Amb, lateral, oral and aboral views.

11–13, hypotype no. 10643e, left Amb, lateral, oral and aboral views.

14–16, hypotype no. 10643f, right InfM, proximal, distal and oral views.

17–19, hypotype no. 10643g, left InfM, proximal, distal and oral views.

Figs. 20–30. *Luidia maculata* Müller & Troschel.   Hypotype no. 10644.

20–22, hypotype no. 10644a, right Adamb, oblique aboral, oblique distal and oblique proximal views.

23, 24, hypotype no. 10644b, left InfM, proximal and distal views.

25–27, hypotype no. 10644c, right InfM, distal, proximal and oral views.

28, 29, hypotype no. 10644d, right Amb, aboral and oral views.

30, hypotype no. 10644e, right Amb, lateral view.

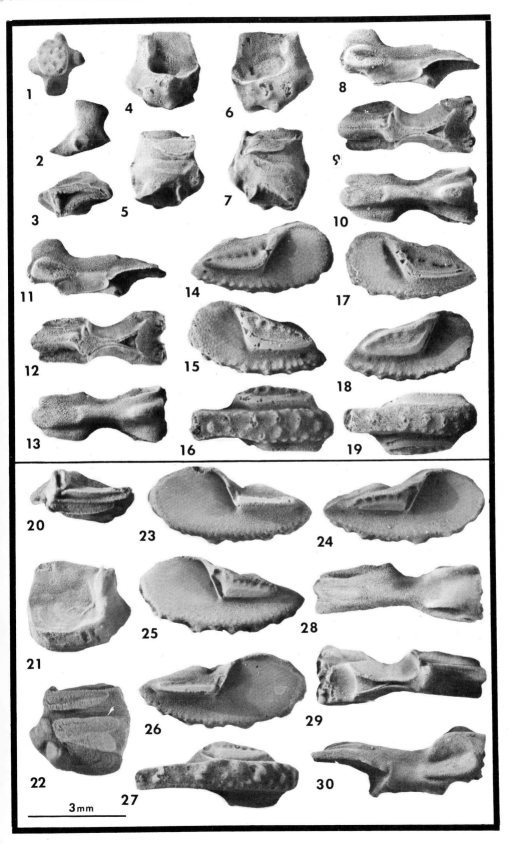

3mm

PLATE 8

Figs. 1–15. *Luidia phragma* Clark.   Hypotype no. 10645.

    1–3, hypotype no. 10645a, right Adamb, oblique proximal, oblique distal and oblique aboral views.

    4, 5, hypotype no. 10645b, left Amb, lateral and aboral views.

    6, 7, hypotype no. 10645c, right Amb, oral and aboral views.

    8, 9, hypotype no. 10645d, left Amb, lateral and oral views.

    10, 11, hypotype no. 10645e, right InfM, oral and distal views.

    12, 13, hypotype no 10645f, right InfM, proximal and oral views.

    14, 15, hypotype no. 10645g, left InfM, proximal and distal views.

Figs. 16–33. *Luidia penangensis* de Loriol.   Hypotype no. 10646.

    16, hypotype no. 10646a, Pax, lateral view.

    17, 20, 21, hypotype no. 10646b, left Adamb, oblique aboral, oblique proximal and oblique distal views.

    18, 19, hypotype no. 10646c, left Adamb, oblique proximal and oblique distal views.

    22, 24, hypotype no. 10646d, left InfM, oral and distal views.

    23, hypotype no. 10646e, right InfM, proximal view.

    25–27, hypotype no. 10646f, left InfM, oral, proximal and distal views.

    28–30, hypotype no. 10646g, right Amb, aboral, lateral and oral views.

    31–33, hypotype no. 10646h, left Amb, aboral, lateral and oral views.

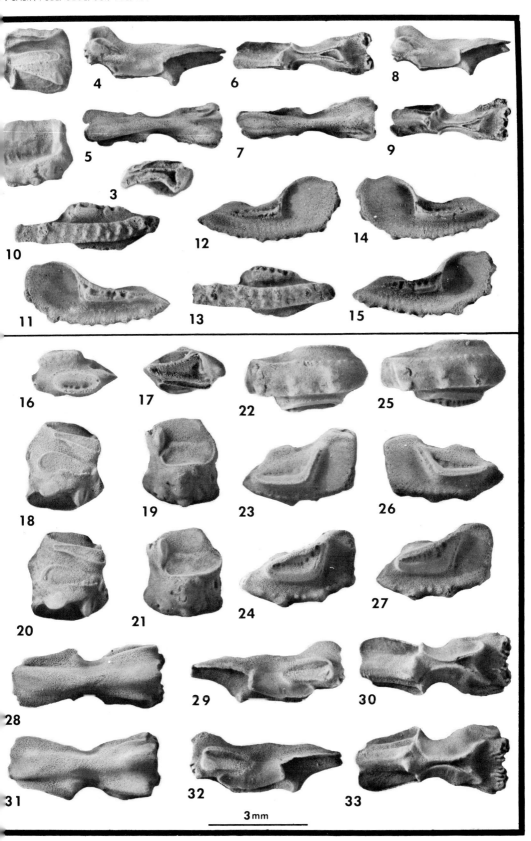

3mm

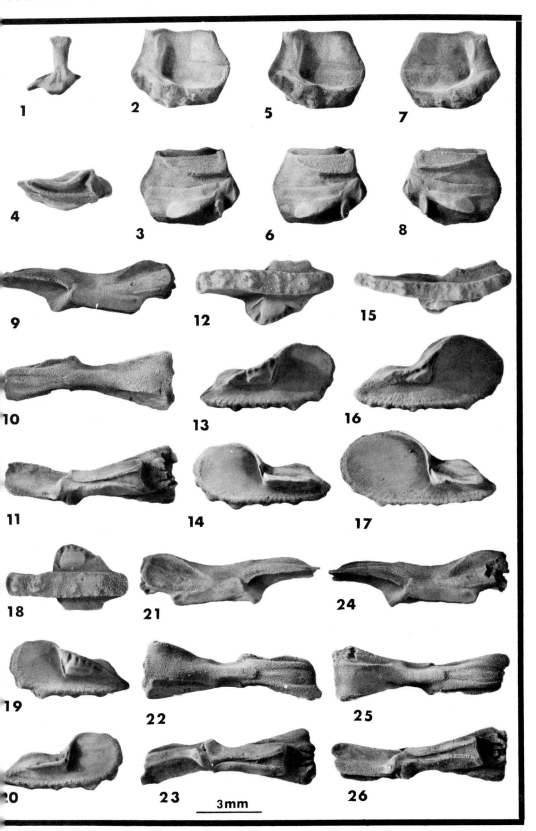

3mm

PLATE 10

Figs. 1–26. *Luidia ciliaris* (Philippi).   Hypotype no. 10648.

1, 2, hypotype no. 10648a, Pax, lateral and aboral views.

3, 8, 9, hypotype no. 10648b, right Adamb, oblique aboral, oblique distal and oblique porximal views.

4, 5, hypotype no. 10648c, left Adamb, oblique distal and oblique proximal views.

6, 7, hypotype no. 10648d, right Adamb, oblique distal and oblique proximal views.

10–12, hypotype no. 10648e, left Amb, aboral, oral and lateral views.

13–15, hypotype no. 10648f, right Amb, aboral, oral and lateral views.

16–18, hypotype no. 10648g, left Amb, aboral, oral and lateral views.

19–20, hypotype no. 10648h, left InfM, distal and proximal views.

21–23, hypotype no. 10648i, right InfM, distal, proximal and oral views.

24–26, hypotype no. 10648j, right InfM, distal, proximal and oral views.

Figs. 27–52. *Luidia elegans* Perrier.   Hypotype no. 10649.

27, 28, hypotype no. 10649a, Pax, aboral and lateral views.

29, 32, hypotype no. 10649b, right Adamb, oblique distal and oblique proximal views.

30–31, hypotype no. 10649c, left Adamb, oblique proximal and oblique aboral views.

33, 34, hypotype no. 10649d, left Adamb, oblique distal and oblique proximal views.

35–37, hypotype no. 10649e, right Amb, aboral, oral and lateral views.

38–40, hypotype no. 10649f, right Amb, aboral, oral and lateral views.

41–43, hypotype no. 10649g, left Amb, lateral, aboral and oral views.

44–46, hypotype no. 10649h, left InfM, distal, proximal and oral views.

47–49, hypotype no. 10649i, right InfM, distal, proximal and oral views.

50–52, hypotype no. 10649j, right InfM, distal, proximal and oral views.

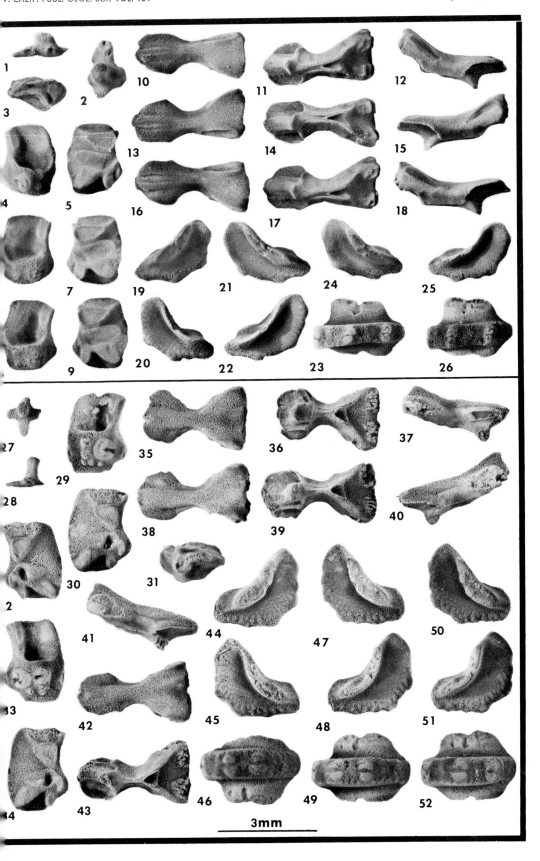

3mm

PLATE 11

Figs. 1–23. *Luidia asthenosoma* Fisher.   Hypotype no. 10650.

1, 2, hypotype no. 10650a, Pax, aboral and lateral views.

3, 4, hypotype no. 10650b, distal left Adamb, oblique distal and oblique proximal views.

5–7, hypotype no. 10650c, left Adamb, oblique aboral, oblique distal and oblique proximal views.

8, 9, hypotype no. 10650d, right Adamb, oblique distal and oblique proximal views.

10–12, hypotype no. 10650e, proximal left Amb, lateral, oral and aboral views.

13, 15, hypotype no. 10650f, proximal right Amb, lateral and aboral views.

14, hypotype no. 10650g, distal left Amb, aboral view.

16, 17, hypotype no. 10650h, distal left InfM, distal and oral views.

18–20, hypotype no. 10650i, right InfM, distal, proximal and oral views.

21–23, hypotype no. 10650j, left InfM, oral, proximal and distal views.

Figs. 24–44. *Luidia neozelanica* Mortensen.   Hypotype no. 10651.

24, 25, hypotype no. 10651a, small Pax, aboral and lateral views.

26, 27, hypotype no. 10651b, large Pax, aboral and lateral views.

28, 31, 32, hypotype no. 10651c, right Adamb, oblique aboral, oblique distal and oblique proximal views.

29, 30, hypotype no. 10651d, left Adamb, oblique distal and oblique proximal views.

33, 39, hypotype no. 10651e, left InfM, distal and oral views.

34, 35, 38, hypotype no. 10651f, left InfM, proximal, distal and oral views.

36, 37, hypotype no. 10651g, right InfM, proximal and distal views.

40, 41, hypotype no. 10651h, right Amb, aboral and oral views.

42–44, hypotype no. 10651i, right Amb, lateral, aboral and oral views.

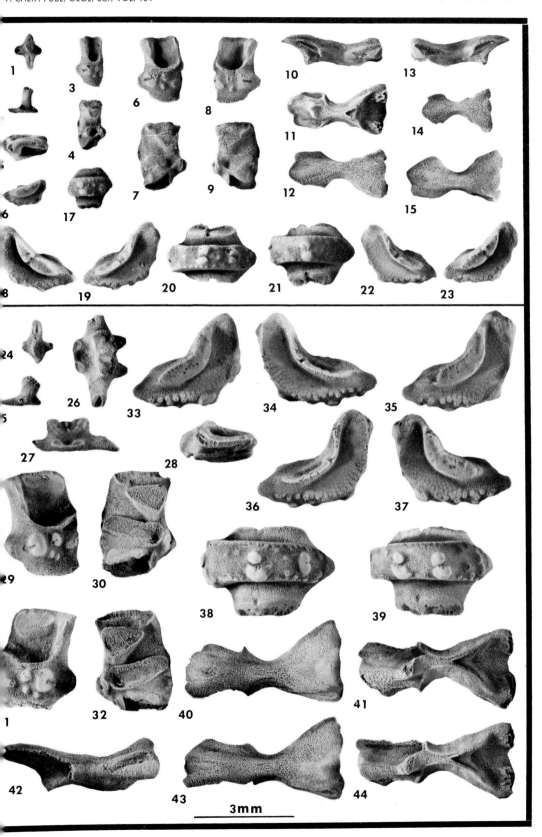

3mm

PLATE 12

Figs. 1–20. *Luidia etchegoinensis*, n. sp.   Holotype no. 10652.

1, Pax, lateral view.

2, 3, holotype no. 10652f, incomplete right Adamb, oblique distal and oblique proximal views.

4–6, holotype no. 10652e, incomplete right Adamb, oblique aboral, oblique distal and oblique proximal views.

7, 8, holotype no. 10652d, incomplete right Adamb, oblique distal and oblique proximal views.

9–11, holotype no. 10652c, adradial Amb fragment, lateral, aboral and oral views.

12, Amb and Adamb in approximate life orientation, proximal view.

13–15, holotype no. 10652a, left InfM, articulation ridge margins chipped, proximal, distal and oral views.

16–18, holotype no. 10652b, left InfM, distal articulation ridge margin chipped, proximal, distanl oral views.

19, oral view of InfM sequence, adradial left, distal down.

20, oblique abradial view of specimen.

Figs. 21-30. *Luidia sanjoaquinensis*, n. sp.   Holotype no. 10653.

21, oral view of abradial fragment of Amb, abradial left.

22, incomplete right Adamb, oblique distal view.

23, lateral view of Amb.

24, holotype no. 10653b, incomplete right Adamb, oblique distal view.

25–27, holotype no. 10653a, right InfM, articulation ridge margins chipped, oral, proximal and distal views.

28, oral view of InfMM with broad paddle-like spines.

29, lateral view of four paxillae.

30, aboral view of paxillae field showing globular granules encrusting paxillae.

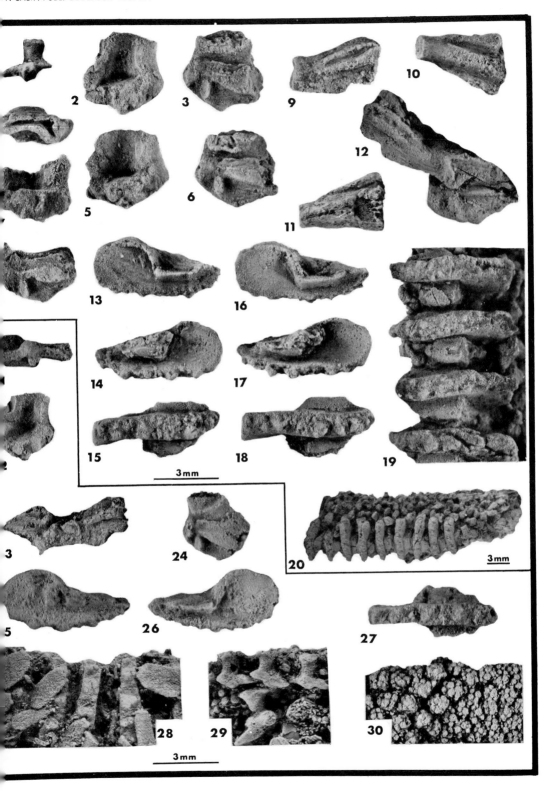

PLATE 13

Figs. 1–22. *Astropecten* sp. Photographs taken from artificial molds.

1, 2, hypotype no. 10656, Pax, lateral views.

3, 5, hypotype no. 10661, left Adambb, oblique proximal views.

4, hypotype no. 10657, left Adamb, oblique proximal view.

6, hypotype no. 10655, right Adamb, oblique proximal view.

7, 8, hypotype no. 10658, right Adambb, oblique distal views.

9, hypotype no. 10655, right Adamb, oblique distal view.

10, 12, hypotype no. 10655, Ambb, aboral and oblique oral views.

11, hypotype no. 10655, Amb, oblique lateral view.

13, hypotype no. 10655, SupM, lateral view.

14, 16, hypotype no. 10656, SupMM, lateral views.

15, hypotype no. 10660, SupM, lateral view.

17, 18, holotype no. 10658, InfMM, oblique lateral views.

19, hypotype no. 10655, InfM, lateral view.

20, hypotype no. 10661, InfM, lateral view.

21, hypotype no. 10658, InfM series, oral view.

22, hypotype no. 10659, aboral view of parts of one arm and disc.

Figs. 23, 24. *Luidia* sp. B. Photographs taken from artificial molds.

23, oral view of arm fragment showing InfMM and Adambb, hypotype no. 10663; 24, oral view of arm fragment showing Ambb, Adambb, and InfMM, hypotype no. 10662.

Figs. 25, 26. *Luidia* sp. C, hypotype no. 10654. 25, oral view of arm fragment; 26, lateral view of InfM.

Figs. 27, 28. *Luidia* sp. A, hypotype USNM 651560. 27, InfM, lateral view; 28, aboral view of arm fragment.

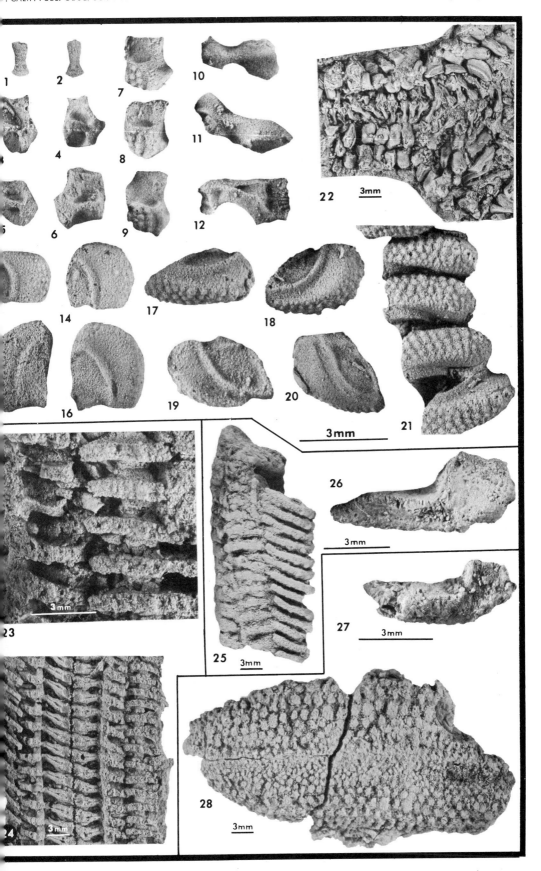

PLATE 14

Figs. 1–12. *Astropecten californicus* Fisher.   Hypotype no. 10666.

1, hypotype no. 10666a, right Adamb, oblique proximal view.

2, hypotype no. 10666b, right Adamb, oblique distal view.

3, hypotype no. 10666c, left Amb, proximal view.

4, hypotype no. 10666d, right Amb, distal view.

5, hypotype no. 10666e, left Amb, oral view.

6, hypotype no. 10666f, right Amb, aboral view.

7, hypotype no. 10666g, distal SupM, lateral view.

8, hypotype no. 10666h, proximal SupM, lateral view.

9, hypotype no. 10666i, medial SupM, aboral view, adradial right.

10, hypotype no. 10666j, medial InfM, lateral view, adradial right.

11, hypotype no. 10666k, proximal InfM, lateral view, adradial left.

12, hypotype no. 10666l, proximal InfM, oral view, adradial right.

Figs. 13–24. *Astropecten regalis* Gray.   Hypotype 10667.

13, hypotype no. 10667a, left Adamb, oblique proximal view.

14, hypotype no. 10667b, left Adamb, oblique proximal view.

15, hypotype no. 10667c, left Adamb, oblique distal view.

16, hypotype no. 10667d, SupM, aboral view.

17, hypotype no. 10667e, SupM, lateral view.

18, hypotype no. 10667f, right Amb, distal view.

19, hypotype no. 10667g, left Amb, proximal view.

20, hypotype no. 10667h, right Amb, oral view.

21, hypotype no. 10667i, right Amb, aboral view.

22, hypotype no. 10667j, InfM, lateral view, adradial left.

23, hypotype no. 10667k, InfM, oral view, adradial left.

24, hypotype no. 10667l, InfM, lateral view, adradial right.

Figs. 25–35. *Astropecten ornatissimus* Fisher.   Hypotype no. 10668.

25, hypotype no. 10668a, left Adamb, oblique proximal view.

26, hypotype no. 10668b, right Adamb, oblique distal view.

27, hypotype no. 10668c, left Amb, aboral view.

28, hypotype no. 10668d, right Amb, proximal view.

29, hypotype no. 10668e, left Amb, distal view.

30, hypotype no. 10668f, SupM, lateral view, adradial left.

31, hypotype no. 10668g, SupM, lateral view, adradial left.

32, hypotype no. 10668h, SupM, aboral view, adradial right.

33, hypotype no. 10668i, InfM, lateral view, adradial left.

34, hypotype no. 10668j, InfM, lateral view, adradial right.

35, hypotype no. 10668k, InfM, oral view, adradial right.

Figs. 36–55. *Astropecten armatus* Gray.   Hypotype no. 10669.

36, hypotype no. 10669a, Pax, lateral view.

37, hypotype no. 10669b, right Adamb, oblique aboral view.

38, hypotype no. 10669c, left Adamb, oblique distal view.

39, hypotype no. 10669d, left Adamb, oblique proximal view.

40, hypotype no. 10669e, right Amb, aboral view.

41, hypotype no. 10669f, left Amb, oral view.

42, 44, hypotype no. 10669g, right Amb, proximal, distal views.

43, hypotype no. 10669h, left Amb, proximal view.

45, hypotype no. 10669i, left Amb, distal view.

46, 47, hypotype no. 10669j, interbrachial InfM, lateral view, oral view, adradial right.

48, hypotype no. 10669k, distal InfM, lateral view, adradial right.

49, hypotype no. 10669l, proximal InfM, lateral view, adradial right.

50, hypotype no. 10669m, proximal InfM, oral view, adradial right.

51, hypotype no. 10669n, proximal SupM, aboral view, adradial right.

52, hypotype no. 10669o, interbrachial SupM, aboral view, adradial right.

53, hypotype no. 10669p, proximal SupM, abradial view.

54, hypotype no. 10669q, proximal SupM, lateral view, adradial left.

55, hypotype no. 10669r, interbrachial SupM, lateral view, adradial left.

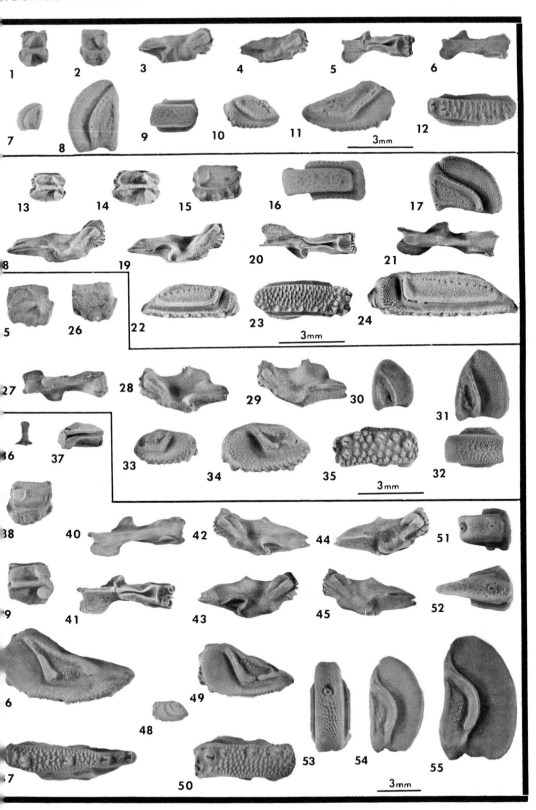

PLATE 15

Figs. 1–11. *Dipsacaster eximius* Fisher.   Hypotype no. 10670.

1, hypotype no. 10670a, left medial Adamb, oblique distal view.

2, hypotype no. 10670b, left proximal Adamb, oblique proximal view.

3, hypotype no. 10670c, left medial Amb, aboral view.

4, hypotype no. 10670d, right proximal Amb, oral view.

5, hypotype no. 10670e, left proximal Amb, distal view.

6, hypotype no. 10670f, medial SupM, aboral view.

7, hypotype no. 10670g, medial SupM, lateral view.

8, hypotype no. 10670h, proximal SupM, lateral view.

9, hypotype no. 10670i, medial InfM, oral view.

10, hypotype no. 10670j, medial InfM, lateral view.

11, hypotype no. 10670k, proximal InfM, lateral view.

Figs. 12–24. *Psilaster pectinatus* (Fisher).   Hypotype no 10671.

12, hypotype no. 10671a, left medial Adamb, oblique proximal view.

13, hypotype no. 10671b, left proximal Adamb, oblique proximal view.

14, hypotype no. 10671c, right proximal Adamb, oblique distal view.

15, hypotype no. 10671d, right medial Amb, distal view.

16, hyoptype no. 10671e, left proximal Amb, oral view.

17, hypotype no. 10671f, right proximal Amb, aboral view.

18, hypotype no. 10671g, proximal InfM, abradial view.

19, hypotype no. 10671h, proximal InfM, lateral view, adradial left.

20, hypotype no. 10671i, medial InfM, lateral view, adradial right.

21, hypotype no. 10671j, medial InfM, oral view, adradial left.

22, hypotype no. 10671k, medial SupM, abradial view.

23, hypotype no. 10671l, medial SupM, lateral view.

24, hypotype no. 10671m, proximal SupM, lateral view.

Figs. 25–36. *Thrissacanthias penicillatus* (Fisher).   Hypotype no. 10672.

25, hypotype no. 10672a, right medial Adamb, oblique proximal view.

26, hypotype no. 10672b, right proximal Adamb, oblique proximal view.

27, hypotype no. 10672c, left proximal Adamb, oblique distal view.

28, hypotype no. 10672d, medial SupM, abradial view.

29, hypotype no. 10672e, interbrachial SupM, abradial view.

30, hypotype no. 10672f, medial SupM, lateral view.

31, hypotype no. 10672g, medial InfM, abradial view.

32, hypotype no. 10672h, interbrachial InfM, abradial view.

33, hypotype no. 10672i, medial InfM, lateral view.

34, hypotype no. 10672j, right medial Amb, lateral view.

35, hypotype no. 10672k, right medial Amb, aboral view.

36, hypotype no. 10672l, right proximal Amb, aboral view.

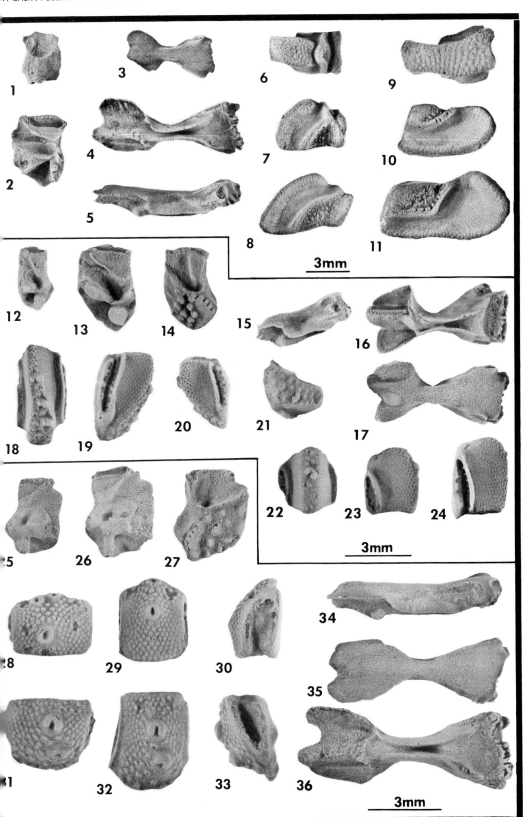

3mm

3mm

3mm

PLATE 16

Figs. 1–15. *Cheiraster gazelle* Studer.   Hypotype no. 10673.

  1, hypotype no. 10673a, arm SupM, abradial view.

  2, hypotype no. 10673b, arm SupM, abradial view.

  3, hypotype no. 10673c, interbrachial SupM, abradial view.

  4, hypotype no. 10673d, arm SupM, aboral view.

  5, hypotype no. 10673e, arm InfM, abradial view.

  6, hypotype no. 10673f, arm InfM, abradial view.

  7, hypotype no. 10673g, interbrachial InfM, abradial view.

  8, hypotype no. 10673h, arm InfM, oral view.

  9, hypotype no. 10673i, right Amb, oral view.

  10, hypotype no. 10673j, right Amb, aboral view.

  11, hypotype no. 10673k, arm Amb, distal view.

  12, hypotype no. 10673l, disc Amb, proximal view.

  13, hypotype no. 10673m, arm Adamb, oblique proximal view.

  14, hypotype no. 10673n, disc Adamb, oblique proximal view.

  15, hypotype no. 10673o, disc Adamb, oblique distal view.

Figs. 16–29. *Pectinaster hylacanthus* Fisher.   Hypotype no. 10674.

  16, hypotype no. 10674a, interbrachial SupM, abradial view.

  17, hypotype no. 10674b, arm SupM, abradial view.

  18, hypotype no. 10674c, arm SupM, abradial view.

  19, hypotype no. 10674d, arm SupM, aboral view.

  20, hypotype no. 10674e, right Adamb, oblique proximal view.

  21, hypotype no. 10674f, left Adamb, oblique distal view.

  22, hypotype no. 10674g, interbrachial InfM, abradial view.

  23, hypotype no. 10674h, arm InfM, abradial view.

  24, hypotype no. 10674i, arm InfM, abradial view.

  25, hypotype no. 10674j, arm InfM, oral view.

  26, hypotype no. 10674k, disc Amb, proximal view.

  27, hypotype no. 10674l, arm Amb, proximal view.

  28, hypotype no. 10674m, left Amb, oral view.

  29, hypotype no. 10674n, right Amb, aboral view.

Figs. 30–44. *Mistia spinosa,* n. gen., n. sp.   Holotype no. 10675.

  30, 31, holotype no. 10675l, left Adamb, oblique distal and oblique proximal views.

  32, 33, holotype no. 10675k, left Adamb, oblique distal and oblique proximal views.

  34, 35, holotype no. 10675a, left Adamb, oblique distal and oblique proximal views.

  36, 37, holotype no. 10675b, incomplete left Adamb, oblique distal and oblique proximal views.

  38, 39, holotype no. 10675i, InfM, oral and abradial views.

  40, 41, holotype no. 10675j, InfM, oral and abradial views.

  42, 43, holotype no. 10675h, InfM, oral and abradial views.

  44, aboral view of holotype.

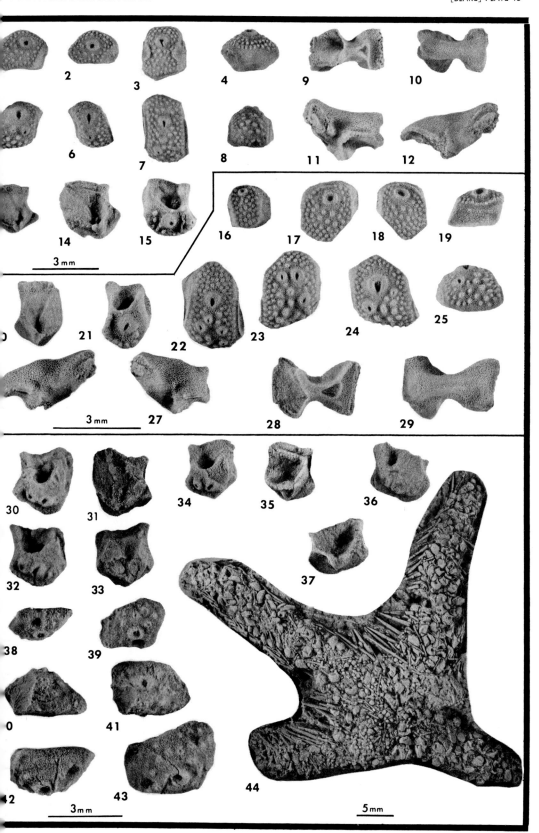

PLATE 17

Figs. 1–21, 35, 36. *Mistia spinosa,* n. gen., n. sp.   Holotype 10675.

1, 2, holotype no. 10675o, Pax, lateral and aboral views.

3, 4, holotype no. 10675n, Pax, lateral and aboral views.

5, 6, holotype no. 10675m, Pax, lateral and aboral views.

7, distal right Amb, oral view.

8, disc Amb, oblique lateral view.

9, 10, holotype no. 10675g, SupM, abradial and aboral views.

11, 12, holotype no. 10675c, SupM, abradial and aboral views.

13–15, holotype no. 10675e, SupM, lateral, abradial, and aboral views.

16–18, holotype no. 10675f, SupM, oblique adradial, abradial, and aboral views.

19–21, holotype no. 10675d, SupM, lateral, abradial, and aboral views.

35, view of portion of two arms and disc of holotype.

36, portion of one arm, showing prominent spines, marginals, Ambb, and Adambb.

Figs. 22–34. Goniasterid sp.A.   Hypotype no. 10676.

22–23, hypotype no. 10676a, Amb, oral and aboral views.

24, hypotype no. 10676b, Adamb, oral view.

25, hypotype no. 10676c, Adamb, oral view.

26, hypotype no. 10676d, Adamb, distal (?) view.

27, hypotype no. 10675e, Adamb, proximal (?) view.

28, hypotype no. 10676f, marginal, outer face.

29, 30, hypotype no. 10676g, marginal, outer, and lateral faces.

31, 32, hypotype no. 10676h, marginal, outer, and lateral faces.

33, 34, hypotype no. 10676i, marginal, outer, and lateral faces.

28, 29, 31, 33, ossicles, showing pits for pedicellariae.

Fig. 37. *Nehelemia delicata,* n. gen., n. sp. Aboral view of holotype no. 10677.

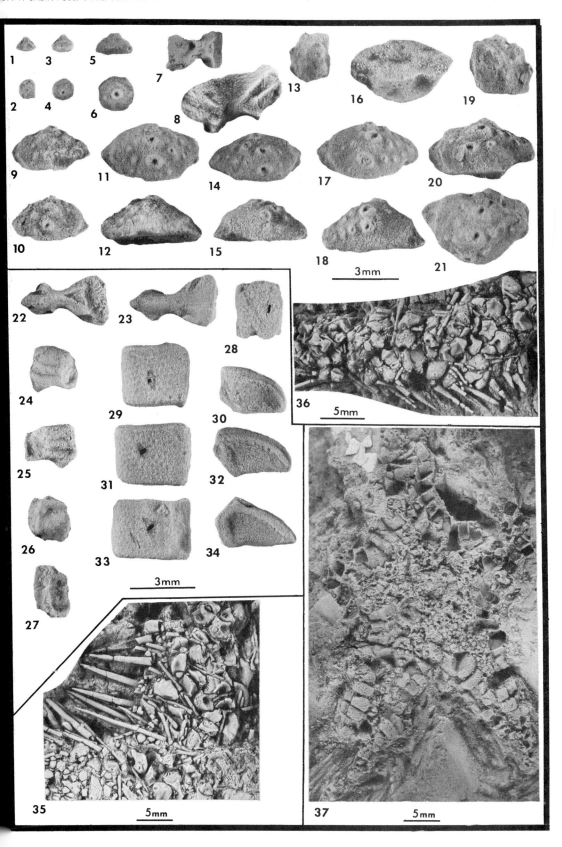

PLATE 18

Figs. 1–13. *Ceramaster leptoceramus* (Fisher). Hypotype no. 10678.

  1, hypotype no. 10678a, Adamb, aboral view, adradial left.

  2, hypotype no. 10678b, Adamb, oral view, adradial left.

  3, hypotype no. 10678c, Adamb, proximal view, adradial left.

  4, hypotype no. 10678d, distal right Amb, oral view.

  5, hypotype no. 10678e, proximal left Amb, oral view.

  6, hypotype no. 10678f, proximal Amb, proximal view.

  7, hypotype no. 10678g, proximal right Amb, aboral view.

  8, hypotype no. 10678h, SupM, lateral view.

  9, hypotype no. 10678i, SupM, aboral view, adradial left.

  10, hypotype no. 10678j, SupM, aboral view, adradial left.

  11, hypotype no. 10678k, InfM, lateral view.

  12, hypotype no. 10678l, InfM, oral view, adradial left.

  13, hypotype no. 10678m, InfM, oral view, adradial left.

Figs. 14–25. *Paragonaster ctenipes hypacanthus* Fisher. Hypotype no. 10679.

  14, hypotype no. 10679a, left Amb, oral view.

  15, hypotype no. 10679b, left Amb, aboral view.

  16, hypotype no. 10679c, right Amb, proximal view.

  17, hypotype no. 10679d, left Adamb, oral view, adradial down.

  18, hypotype no. 10679e, right Adamb, oral view, adradial down.

  19, hypotype no. 10679f, proximal SupM, aboral view.

  20, hypotype no. 10679g, proximal SupM, lateral view.

  21, hypotype no. 10679h, interbrachial SupM, lateral view.

  22, hypotype no. 10679i, carinal, aboral view.

  23, hypotype no. 10679j, interbrachial InfM, oral view, adradial right.

  24, hypotype no. 10679k, proximal InfM, lateral view.

  25, hypotype no. 10679l, interbrachial InfM, lateral view.

Figs. 26–37. *Mediaster aequalis* Stimpson. Hypotype no. 10680.

  26, hypotype no. 10680a, Adamb, aboral view, adradial left.

  27, hypotype no. 10680b, Adamb, oral view, adradial left.

  28, hypotype no. 10680c, left Adamb, proximal view.

  29, hypotype no. 10680d, right Amb, proximal view. Arrows point to adradial (ad) and abradial (ab) apophyses.

  30, hypotype no. 10680e, left Amb, oral view.

  31, hypotype no. 10680f, left Amb, aboral view.

  32, hypotype no. 10680g, SupM, aboral view.

  33, hypotype no. 10680h, SupM, lateral view.

  34, hypotype no. 10680i, SupM, lateral view.

  35, hypotype no. 10680j, InfM, oral view.

  36, hypotype no. 10680k, InfM, lateral view.

  37, hypotype no. 10680l, InfM, lateral view.

Figs. 38–48. *Nehalemia delicata*, n. gen., n. sp. Holotype no. 10677.

  38, holotype no. 10677a, right Adamb, oral view, adradial up.

  39, holotype no. 10677b, marginal, lateral view.

  40, holotype no. 10677c, marginal, lateral view.

  41, holotype no. 10677d, marginal, lateral view.

  42, holotype no. 10677e, marginal, lateral view.

  43, holotype no. 10677f, marginal, outer face, adradial right.

  44, holotype no. 10677g, marginal, outer face, adradial right.

  45, holotype no. 10677h, marginal, outer face, adradial right.

  46, holotype no. 10677i, marginal, outer face, adradial right.

  47, holotype no. 10677j, marginals, outer faces, adradial right.

  48, holotype no. 10677, portion of disc and one arm, showing marginals and aboral disc ossicles.

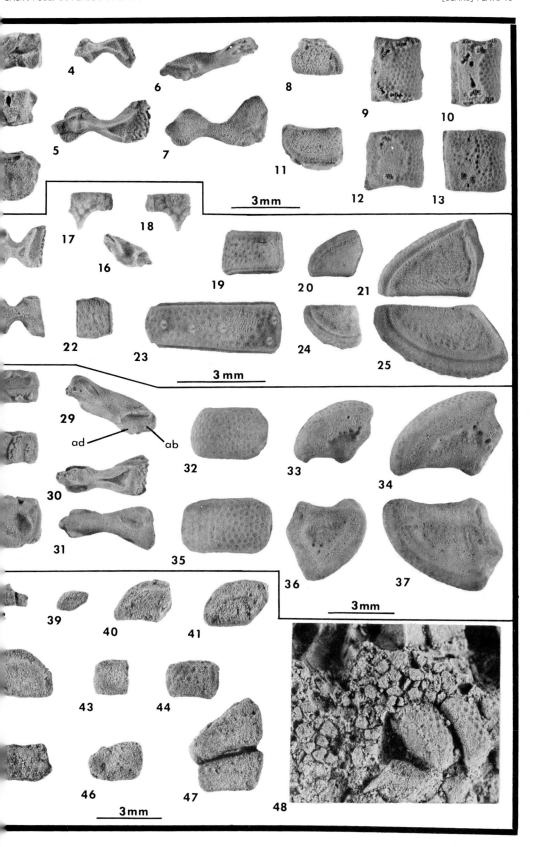

PLATE 19

Figs. 1–16. *Sucia suavis*, n. gen., n. sp.   Holotype no. 10665.

1, right Adamb, oral view, adradial down.

2, distal SupM, lateral view.

3, three carinals and associated arm ossicles, adradial view.

4, 12, 15, lateral views of arm sections.

5, oral view of interbrachial and arm InfM, adradial up, distal right.

6, 7, proximal view of arm SupMM and InfMM.

8, "adradial" view of interbrachial SupMM and InfMM.

9, aboral view of holotype.

10, oral view of holotype.

11, 14, lateral views of interbrachial SupMM and InfMM.

13, 16, cross sections of arms, showing marginals, carinals, and partially obscured Ambb and Adambb.

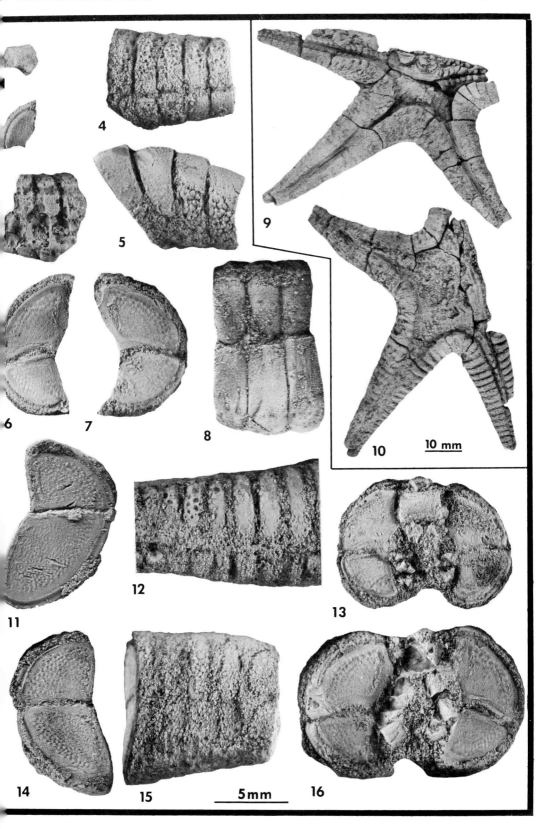

4

5

6

7

8

9

10

10 mm

11

12

13

14

15

5mm

16

ISBN: 0-520-09472-7

Randall Library – UNCW

QE783.A7 B55   NXWW

Blake / Ossicle morphology of some recent asteroid

3049001938824